ECOLOGICAL ASPECTS OF TOXICITY TESTING OF OILS AND DISPERSANTS

*Proceedings of a Workshop on the
Toxicity Testing of Oils and Dispersants
held at the Institute of Petroleum, London*

ECOLOGICAL ASPECTS OF TOXICITY TESTING OF OILS AND DISPERSANTS

Edited by

L. R. BEYNON and E. B. COWELL

British Petroleum Co. Ltd,
London, UK

A HALSTED PRESS BOOK

JOHN WILEY & SONS

New York — Toronto

PUBLISHED IN THE U.S.A. AND CANADA BY
HALSTED PRESS
A DIVISION OF JOHN WILEY & SONS INC., NEW YORK

The symbol I.P. on this book implies that the text has been officially accepted as authoritative by the Institute of Petroleum, Great Britain

Library of Congress Cataloging in Publication Data
Main entry under title:

Ecological aspects of toxicity testing of oils and
  dispersants.

  "A Halsted Press book."
  "Proceedings of a workshop on the toxicity testing
of oils and dispersants held at the Institute of
Petroleum, London."
  1. Oil spills and wildlife—Congresses. 2. Petroleum—
Toxicology—Congresses. 3. Dispersing agents—
Toxicology—Congresses. I. Beynon, L. R., ed.
II. Cowell, E. B., ed. III. Institute of Petroleum,
London.
QH91.8.04E25   574.2'4   74–12044
ISBN 0–470–07190–7

WITH 26 ILLUSTRATIONS AND 13 TABLES

© APPLIED SCIENCE PUBLISHERS LTD 1974

PRINTED IN GREAT BRITAIN BY GALLIARD (PRINTERS) LTD GREAT YARMOUTH

Contents

Preface

A Workshop on Toxicity Testing of Oils and Dispersants was held at the Institute of Petroleum, 61 New Cavendish Street, London. The Workshop was convened by the Working Party on Dispersants of the Institute of Petroleum's Sea Pollution Committee. Representatives from various European laboratories which play a prominent part in the toxicity-testing field attended the Workshop, contributing formal papers and taking part in the discussion.

In preparing the proceedings of the Workshop the Editors decided that it was worthwhile presenting the bulk of the discussion verbatim, and only minor editorial changes and omissions have been made.

Following the Workshop, a paper on 'Toxicity Testing of Oils and Dispersants: A European View' was prepared by K. W. Wilson, E. B. Cowell and L. R. Beynon, taking into account the views expressed at the Workshop. This paper was also presented at the Conference on the Prevention and Control of Oil Spills, sponsored by the American Petroleum Institute, the Environmental Protection Agency and the US Coastguard, which was held at Washington DC. For completeness this paper is included in these proceedings. The Editors are grateful to the sponsors of the Washington Conference for permission to reproduce the paper here.

L.R.B.
E.B.C.

Acknowledgements

Not all the laboratories invited to send representatives to the Workshop were able to be present.

Representatives of the following organisations made valuable contributions:

The Institute of Petroleum's Sea Pollution Committee.
The Institute of Petroleum's Working Party on Dispersants.
Kristineberg Zoologiska Station, Sweden.
Institute for Marine Environmental Research, England.
Fisheries Resources Division, FAO, Italy.
Berridge International Consultants, England.
Swedish Coastguard Service, Sweden.
Field Studies Council, Oil Pollution Research Unit, Orielton Field Centre, Pembroke, Wales.
Department of Zoology, University of Cork, Eire.
Fisheries Laboratory, Ministry of Agriculture, Fisheries and Food, England.
Marine Laboratory, Department of Agriculture and Fisheries for Scotland, Scotland.
Stichting CONCAWE, Netherlands.
Department of Biology, University of Strathclyde, Scotland.
Askö Laboratory of the University of Stockholm, Sweden.
International Tanker Owners' Pollution Federation, England.
Biologische Anstalt Helgoland, Helgoland, Germany.
Marine Biological Association of the United Kingdom, England.
National Environmental Protection Board, Sweden.
Admiralty Oil Laboratory, England.
Station Marine d'Endoume, University of Aix-en-Provence, France.
Warren Spring Laboratory, Department of Trade and Industry, England.
Institute of Hygiene and Epidemiology, Belgium.
Department of Applied Biology, Chelsea College, England (representing NATO).
British Industrial Biological Research Association, England.

1
Introduction

L. R. BEYNON

(*British Petroleum Co. Ltd, Chairman of the Institute of Petroleum's Working Party on Dispersants*)

The IP, largely through its Co-ordinating Committee for the Prevention of Sea Pollution, has been deeply involved in oil pollution problems, causes, effects and clean-up for many years. Overall I think that the IP (in association with UK Government Departments) can fairly claim to have played a leading role in such matters—both before the famous accident to the *Torrey Canyon* and more particularly since that time. One of the ways in which the IP has carried out its work in the oil pollution field is through specialised working groups. One of these has been concerned with 'Dispersants and their Applications' and, as Chairman of this working group, it is my privilege to act as Chairman of this Workshop over the next two days.

The effects on marine life, and particularly on littoral communities, of both oil spillages and clean-up methods have been the subject of worldwide concern and study. Once again, the year of the *Torrey Canyon*, 1967, was a turning-point in these matters because it focused the attention of the world on the problems with which all maritime nations could be faced. But although there has been a common concern about the effects of oil and of oil spill dispersants, for example, there has been little agreement regarding how such effects should be measured and predicted. This divergence of opinion has been particularly apparent between European scientists and those employed in such studies in North America.

This Workshop was convened to examine the methods of toxicity testing commonly practised in leading European marine biological laboratories—in terms of the toxicity of oil, the toxicity of dispersants, and of mixtures of the two. Although this may sound a very simple matter to the layman, I am sure that everyone will agree that, in practice, it is a very complex subject. 'Oil' is a simple term but covers a vast number of variables, and there is also a wide variation between the properties of the many dispersants which are now commercially available for dealing with oil spills.

By and large there are two main reasons for carrying out toxicity tests: firstly, to produce comparative rankings of toxicity in standard form, to enable a choice to be made between commercially available dispersants; and secondly, to enable ecological predictions to be made of the effects of spillages and of dispersant usage. We are being very ambitious in trying to consider in detail the methods of testing appropriate to these two approaches in less than

1

two days. Hopefully, however, our discussions will result in at least a clari-
fication of why approaches taken to toxicity testing vary and how the results
of such tests should be interpreted and used. Perhaps, one day, we shall have
accepted Codes of Practice for toxicity testing of hydrocarbons and dispersants. In the meantime, there is a very definite need to arrive at some broad
consensus of opinion regarding how such tests should be carried out and to
document the views of various authorities.

A paper entitled 'Toxicity Testing of Oils and Dispersants: A European
View' will be presented at the Conference on the Prevention and Control of
Oil Spills, to be held in Washington in 1973 under the joint sponsorship of the
American Petroleum Institute, the Environmental Protection Agency and the
US Coastguard. In preparing this paper, the views and data presented at this
Workshop will be extensively drawn upon and the paper will be reprinted in
full in the proceedings (p. 129).

2
Oils and Dispersants: Chemical Considerations

E. N. DODD

(Berridge International Consultants, Southampton, England)

I. OIL

The toxicity of oil alone is a matter which is attended by some degree of ambiguity on account of the difficulty of defining the nature of the oil. While crude oil, containing molecular species ranging from, say, C_4 to C_{40+} may well be toxic to marine organisms, the likelihood of this unweathered material coming into contact with marine life is dubious in view of the rapid loss of the volatile constituents on exposure to the air. For example, it has been shown[1] that under conditions of normal atmospheric turbulence on open water, C_5 and C_6 will disappear in about one hour, while after five hours the species up to C_8 will have evaporated. Within 24 hours 30–40% of the crude may have disappeared, but the possibility must also be borne in mind that during this time some of the light ends may well have dissolved in the water and so be brought into contact with marine life in areas not directly contaminated by the visible oil. These considerations underline the difficulties attending the design of a toxicity test; if a suspension in water of the whole crude is used, the lower molecular weight hydrocarbons are likely to be present in much greater proportion than they are in the crude itself, while if a topped or weathered crude is used, the amount of hydrocarbon passing into the water is likely to be minimal on account of the extremely low solubilities of the higher molecular weight species.

With regard to the toxicity significance of the various fractions of crude, the low-boiling saturated hydrocarbons have been shown to produce anaesthesia and narcosis at low concentrations, and at higher concentrations cell damage to a wide variety of lower animals, in particular to the larval and other young forms of marine life.[2] Higher-boiling saturated hydrocarbons may not be directly toxic, although it is suggested that they may interfere with nutrition. The aromatic hydrocarbons abundant in petroleum form its most dangerous fraction. The low-boiling aromatics—benzene, toluene, xylenes, etc.—represent the acute toxic hazard in the crude, while the higher molecular weight polynuclear species may well be of significance in their long-term effects, in particular, the possibility of their persistence allowing them to pass up the food chain. The assessment of the toxicity of these latter compounds will clearly require a totally different approach to that of the more rapid, acute effects produced by the lower aromatics.

Evidence on oil toxicity from field observations will, as would be expected, be variable. Oil coming ashore from spillages such as that from *Torrey Canyon* showed little evidence of harming marine life when free from dispersant, and this can undoubtedly be associated with the fact that this oil had weathered for several days on passage to the coast and so lost a large proportion of its more volatile constituents. Analyses of samples showed that, in general, molecular species up to C_{10} were absent, while others somewhat less volatile had nevertheless been diminished in concentration.[1]

Where oil contamination is very heavy, there will, of course, always be the possibility of shellfish, for example, becoming smothered, with consequent interference with respiration and browsing activity. On the other hand, that the oil alone is not necessarily toxic is shown by the observation that limpets can ingest the oil and pass it through the gut.[3] Similar observations have been recorded with respect to mussels. In general, the deleterious effect of oil on marine life would appear to be physical rather than chemical; for example, it has been reported[4] that oil contamination of the gill filaments of fish prevents the exchange of gases and so causes anoxia.

Comment should be made on the 'solubility' of hydrocarbons in water, in view of its relevance to the design of a toxicity test. Figures obtained by vigorous shaking with water, standing for 24 hours and filtering before estimation, ranged from 11 ppm for hexane to 0·15 ppm for decane in the case of sea water, while the same method gave slightly higher figures for fresh water.[5] Other workers have reported considerably higher figures,[6] and it is clear that great care is needed to ensure that the determination does not include finely dispersed oil. This was illustrated by shaking 10 ml of oil with 2·5 litres of sea water, allowing to stand, and withdrawing samples at intervals. The oil contents measured ranged from 31 ppm after 15 minutes to 4·6 ppm after 1 day, 2·7 ppm after 2 days, and 0·6 ppm after 147 days. These figures were obtained with Kuwait crude, and could well be different for other crudes in view of varying contents of the respective hydrocarbons. Data obtained in the field sometimes give higher figures, but these may well be due to small traces of 'natural' surfactant in the water holding more oil in suspension than normal, or even to the inclusion of oil of recent biogenic origin.

II. DISPERSANTS

(a) General Characteristics

Oil spill dispersants normally consist of a surface-active agent (surfactant) mixed with a hydrocarbon solvent and sometimes a stabiliser. Thus, BP 1002 consisted of 12% of alkylphenol ethoxylate dissolved in a 60–70% aromatic hydrocarbon solvent, with the addition of 3% fatty acid ethanolamide as stabiliser.[7] While the majority of dispersants documented appear to be based on a petroleum solvent, some consist of surfactant and a water-miscible solvent such as isopropyl alcohol. Dispersol OS is an example of this type, in which 10% of a non-ionic polyethanoxy compound is dissolved in iso-propanol, while two varieties of Corexit are available in which the same surfactant—a non-ionic polyhydric alcohol ester of fatty acids—is mixed either with isopropyl alcohol (7664) or an aliphatic hydrocarbon solvent (8666).

As is well known, surfactants can be divided into three groups, anionic, cationic and non-ionic. The anionic types form the largest group, being largely used in household detergents, and mostly consist of alkali salts of organic sulphates or sulphonates in which the anion is surface-active. Earlier dispersant formulations made use of these, notably alkyl benzene sulphonates, and some presently used contain a proportion of such materials. Cationic types in which the cation is surface-active are mainly based on alkylamines or pyridine structures and form the smallest group. They are not widely employed in the formulation of dispersants because their surface activity is lower than that of the anionic or non-ionic types.

The most widely used non-ionic surfactants are those that are obtained by condensing long-chain alkyl alcohols and alkyl phenols with ethylene oxide. The products obtained by this condensation do not correspond to a single chemical compound, and are usually described by the average number of molecules of ethylene oxide per molecule of alcohol and alkyl phenol. Because the number of variations of both lipophilic and hydrophilic group-ings is considerable, it is possible to make these compounds suit particular applications, both by varying the size of the hydrophilic group and the nature of the lipophilic. Owing to the presence within the molecule of both hydro-philic and lipophilic characteristics, the surfactant can orient itself in the oil/water interface with the lipophilic group(s) in the oil and the hydrophilic in the water, so producing fine oil droplet formation when provided with mixing energy. A further factor of importance is the prevention of coalescence of the oil droplets after they have been formed, i.e. the 'stability' of the oil-in-water emulsion, and this presumably is associated with the balance between the lipophilic and hydrophilic characteristics, so that the former is sufficiently strong to resist removal of the surfactant molecule from the oil droplet dispersed in the water phase. This stability effect would also help to minimise the adhesion of dispersed oil droplets to surfaces, while the continuing maintenance of a vastly increased oil/water interface should increase the rate of biodegradation of the oil.

Of the environmental parameters to be considered in assessing the efficiency of a surfactant, that of temperature is important; evidence to date suggests that the dosage required increases as temperature decreases, roughly in accordance with the temperature dependence of chemical reaction rates.

The effect of suspended solids in the open sea should not be significant, but in coastal, estuarine and river waters the much higher concentrations of suspended material may, by providing competing surfaces for distribution of the surfactant, influence its efficiency.

As indicated above, dispersants may employ either a water-miscible solvent with a predominantly water-compatible surfactant, or a hydrocarbon-base solvent with an oil-compatible surfactant. In the first case, small oil droplets must be produced when the contact is at a maximum, and the prompt application of mixing energy is particularly important here. A common and effective method of achieving this is that of using an eductor where the dispersant can be metered into the system. With petroleum-base dispersants, however, the undiluted material needs to be sprayed on to the oil layer for effective dispersion, and the function of the solvent is, of course, to facilitate the rapid distribution of the surfactant throughout the oil. If mixed with water prior to contact, a solvent-in-water dispersion is formed and it is difficult for the surfactant to transfer from its thermodynamically stable location at the solvent/water interface to the oil/water interface.

A very desirable characteristic of an oil dispersant would be the ability to invert the intractable water-in-oil emulsions, so transforming the oil into the disperse phase when dilution and distribution effects could become operable. The chemical factors in crude oil responsible for the rapid formation and continuing stability of water-in-oil emulsions are being studied, and it is to be hoped that information may be forthcoming which would assist the design of surfactants capable of bringing about the desired phase inversion.

(b) Toxicity

The relationship between the toxicity and chemical structure of the group of surfactants based on ethylene oxide condensation has been attributed to the degree of polymerisation of the polyoxyethylenic chain.[8] The intrinsic toxicity is attributed to the alkyl phenolic radical and a diluent action to the polyoxyethylenic chain, so that the longer the latter, the weaker the toxicity. It has been shown that no toxicity arises from polyoxyethylene compounds alone.

This effect is illustrated by the data shown in Table 1.

TABLE I

Compound	No. of ethylene oxide molecules	Lethal concentration at 6 hours (mg/litre)
Polyoxyethylenic ether	(100%)	>2500
Oxyethylated castor oil	45	>2500
Oxyethylated oleic acid	21	650
Oxyethylated octyl phenol	17	410
Oxyethylated octyl phenol	8·5	11·5
Oxyethylated nonyl phenol	10	5·9–7·5
Oxyethylated tridecyclic acid	5	8·5
Oxyethylated laurylic acid	4	5·2

While the toxicity characteristics of the surfactants alone are important, it is of greater relevance to consider those of the formulated dispersant. A list of some 30 dispersants in current use is given in Beynon's 1970 paper,[9] and the great difference in toxicity between the earlier types used at the time of *Torrey Canyon* and those which have been developed subsequently is at once apparent. The 48-hour LC_{50} values in respect of *Crangon crangon* range from 4 to over 3300 ppm, and examination of the relationship between toxicity and composition shows that this resides primarily in the nature of the solvent. For example, BP 1002 with an LC_{50} value of 6 employs a hydrocarbon solvent containing 60–70% aromatics, while BP 1100 with an LC_{50} value of more than 3300 is based on a non-aromatic solvent. Similarly, other dispersants such as Corexit and Dispersol, with LC_{50} values in excess of 3300 are based either on aliphatic hydrocarbon solvents or on a water-miscible vehicle such as isopropyl alcohol.

(c) Specification

The design of a specification to govern the supply of oil spill dispersants has been discussed by Jeffery[10] and its requirements are shown at Annex A. The link between toxicity and aromatic content of the solvent is recognised by limiting the latter to 3%. The fire hazard is also considered to be of importance, especially in relation to storage in containers where slight leakage is possible. A fire hazard could also exist on board ship during use and in confined areas of water, so that the flash point has been limited to 61°C. This would, of course, prohibit the use of isopropanol (flash point 12°C) unless used only as a minor constituent of the solvent. Halogenated materials such as carbon tetrachloride or trichlorethylene are also prohibited, but these do not appear to be common in dispersant formulations.

REFERENCES

1. 'The Weathering of *Torrey Canyon* Crude Oil', AOL Technical Note No. 32, 1967.
2. Effects of detergents and oils on the cell membrane, *Field Study Council* (suppl.), 2, 131 (1968).
3. '*Torrey Canyon:* Pollution and Marine Life', Report of Plymouth Laboratory, MBA, 1968.
4. Canevari, G. P., 'The role of chemical dispersants in oil cleanup', in Hoult, D. P. (ed.), *Oil in the Sea*, Plenum, New York, 1969, p. 29.
5. 'Dispersion and Solubility of Crude Oil in Water', Admiralty Materials Laboratory Report No. 17, 1971.
6. 'Effects of Natural Factors on the Fate of Oil at Sea', Section VII, MOD (Navy), 1971.
7. 'Oil Spill Dispersants Product Data', Edison, NJ, Water Quality Laboratory, 1971.
8. Marchetti, R., A critical review of the effect of synthetic detergents on aquatic life, *Stud. Rev. Gen. Fish Council Medit.*, 26, 1–32 (1965).
9. Beynon, L. R., 'Oil Spill Dispersants', Workshop on Oil Spill Clean-up, 1970.
10. Jeffery, P. G., 'Dispersants for Oil Spill Clean-up Operations', Warren Spring Laboratory, LR 162(PC), 1971.

ANNEX A

Specification

1. *General*

This specification relates to the supply of detergents, dispersants, emulsifiers, solvent emulsifiers and similar materials required for use in beach-cleaning operations and for oil dispersal at sea. These materials are handled in the liquid phase, and for this reason the material offered against this specification should, under normal operating conditions, contain no solid material, no suspended matter, and no additional liquid phases. It should be non-corrosive to the storage containers and should not contain substances that are normally considered to be toxic to man.

2. Prohibited Ingredients

The dispersant supplied shall not contain benzene, carbon tetrachloride or other chlorinated hydrocarbons, phenol, cresols, caustic alkali, or free mineral acid.

3. Flash Point

The flash point of the dispersant, as determined by Pensky Marten closed-cup method (IP 34/67), shall be not less than 142°F (61°C).

4. Cloud Point

The cloud point of the dispersant (IP 219/67) shall be not more than 14°F (-10°C).

5. Viscosity

The viscosity of the dispersant ready for use (IP 71/66), as measured at 32°F (0°C), shall be not more than 50 centistokes.

6. Hydrocarbon Solvent

The hydrocarbon solvent, if one is used in the manufacture of the dispersant, shall be low in aromatic hydrocarbons, with an upper limit of 3% total aromatic hydrocarbons, as determined by absorption spectroscopy.

7. Containers

The dispersant shall be supplied in standard steel drums of 40–50 gallon capacity.*

8. Labelling

All containers shall be clearly labelled with the name of the supplier, the name, number or other mark used to identify the material, the batch number and the flash point.

9. Shelf Life

The dispersant, in sealed steel drums as supplied, shall have a shelf life of not less than five years.

10. Sampling

A sample of not less than 1 gallon of the material to be supplied against this specification shall be submitted to the Director of Warren Spring Laboratory, PO Box 20, Stevenage, Herts. The supplier shall undertake that all supplies of material to be delivered against this specification shall conform in every way (except in relation to the sample container) to that of the sample supplied to the Laboratory. The Director of Warren Spring Laboratory will arrange for the dispersant sample supplied to be tested for efficiency in promoting the emulsification and dispersal of oil, the methods of such testing to be in accordance with the standard practice at the Laboratory for the testing of such dispersants.

The supplier is to state clearly whether the dispersant is suitable for use in situations where low toxicity to marine life is necessary. These low-toxicity dispersants shall have an LC_{50} to brown shrimps (*Crangon crangon*) greater than 3300 ppm in a 48-hour exposure test.

The Director of Warren Spring Laboratory will arrange for part of the sample supplied to be examined for its toxicity by the Burnham-on-Crouch Laboratory of the Ministry of Agriculture, Fisheries and Food, according to standard procedures in use at that laboratory.

* Special arrangements may be made for bulk deliveries.

11. *Composition and Manufacturer's Specification*

Each supplier will be required to submit to the Director of Warren Spring Laboratory details of the composition and specification used in the preparation of his product. This information will be submitted in confidence, and will not be disclosed by the Director of Warren Spring Laboratory. The manufacturer will also be required to submit details of the scale of production that can be undertaken, and if necessary maintained, to meet exceptional, urgent or emergency requirements for such materials.

12. *Variation of Specification*

All material supplied shall conform to this specification. In the event of a manufacturer being unable to maintain supplies of material under emergency conditions, full details of changes that will be required to meet the heavy demand (for example, changes in solvent composition) must be notified in advance. Such 'modified' materials may not be supplied without special authority, and the containers holding such modified material must be clearly marked with the letter M, both in front of and behind the letters and/or number identifying the product. (Thus, for example, a product normally identified by the letters ABC followed by the numbers 123 would, if supplied in a modified form, be labelled MABC 123M.)

3

Toxicity Testing for Ranking Oils and Oil Dispersants

K. W. WILSON

(Ministry of Agriculture, Fisheries and Food, Fisheries Laboratory, Burnham-on-Crouch, Essex, England)

INTRODUCTION

In 1962 the Warren Spring Laboratory of the Department of Trade and Industry reported that the use of solvent emulsifiers was an effective method of dispersing oil spilt at sea, and that if applied correctly they were unlikely to present any risk to fishery interests. This was largely borne out by events following the extensive use of solvent emulsifiers (= dispersants) to clear up the oil from the *Torrey Canyon*. Local fish stocks were not affected directly to any great extent and they have shown no subsequent changes that can be attributed to the use of dispersants.

Littoral plants and animals received heavy treatment with oil and dispersants and suffered heavy mortality.[11,12,18] Laboratory investigations into the effects of dispersants demonstrated not only that short exposures to low concentrations were acutely toxic to a wide variety of marine species,[3,4,17] but also that they could induce detrimental effects which were not apparent until several months after exposure.[14,21] As a direct result of these and similar studies, efforts to develop less toxic dispersants were intensified.

Since 1967 the Fisheries Laboratory at Burnham-on-Crouch has co-operated with the Warren Spring Laboratory in assessing new dispersant formulations for their efficiency in dispersing oil at sea and on beaches and for their toxicity towards marine animals. Toxicity is measured by a standard technique which aims primarily at producing a rank order of toxicity of the dispersants. This arrangement has enabled large numbers of chemicals to be screened, and those that have represented a significant advance in terms of efficiency and toxicity have been tested more thoroughly.

This account describes some of the factors to be considered in establishing a standard technique of toxicity testing for ranking oil dispersants. The value of this test for assessing toxic effects in the field will be discussed elsewhere.

THE COLLECTION AND TREATMENT OF DATA

In his recent review on the measurement of pollutant toxicity to fish, Sprague[19] reports that, broadly speaking, there are two procedures in current use. In the

11

first, mortalities in several concentrations of the substance under test are recorded only at fixed time intervals, usually 24, 48 or 96 hours, and the concentration lethal to half the animals after these times is interpolated from these data. In the second approach, the time to death of each individual animal is recorded and the time taken to obtain 50% mortality is calculated for each concentration. This latter procedure is recommended since it provides considerably more information from an equal number of animals.

When the percentages of animals dying in a succession of equal time periods at any lethal concentration of poison are plotted against time, the resulting figure represents a normal, but skewed, distribution. If the cumulative percentages are plotted against time they lie approximately on a sigmoid curve. The use of standard statistical techniques to describe this data is valid only if the distributions of survival times, or the transformed survival times,

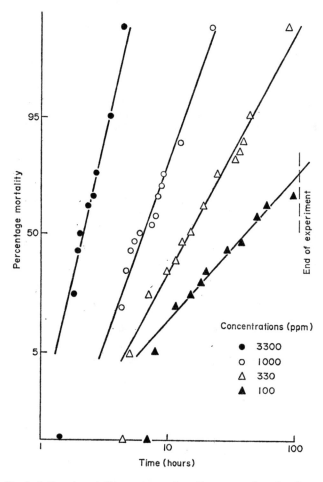

FIG. 3.1. Typical time/mortality curves of a dispersant for the brown shrimp *Crangon crangon*.

are normal. One method of examining the normality of the transformed data
is to examine them mathematically for goodness of fit. A more straightforward
technique is to plot the data graphically, expressing percentage mortality as
probits; the transformed survival times then lie on a straight line if they are
normally distributed. The suitability of a logarithmic time/probit transformation for time/mortality responses has been shown on numerous occasions[1,8]
since the detailed descriptions of Bliss.[2] Litchfield[10] describes how the basic
statistics of the line can be obtained directly by plotting the original data on
log-probability paper. These statistics are given as:

$$\text{mean, } \bar{x} = \text{time at 50\% mortality (ET}_{50})$$

and variance is given by its equivalent S, the slope function, where

$$S = \frac{(\text{ET}_{84}/\text{ET}_{50}) + (\text{ET}_{50}/\text{ET}_{16})}{2}$$

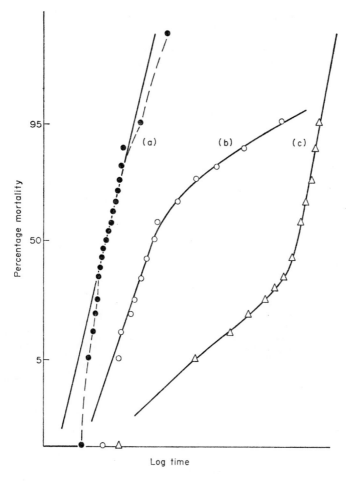

FIG. 3.2. Time/mortality curves for *Crangon crangon*, showing deviations from
the linear relationship (see text for explanation).

where ET_{16}, ET_{84} are the times for 16% and 84% mortality respectively. Thus it is possible to describe fully the response of the population with these terms. Considerably more accuracy is gained by measuring the average rather than a minimum or maximum response, since the extremes of the distribution are so dependent on the size of the test population.

Typical examples of transformed time/mortality curves for a dispersant are shown in Figure 3.1. Deviations from the straight-line relationship are encountered frequently. Apparent changes of slope at the extreme ends of the distributions are especially common (Fig. 3.2a) and are probably due to the small numbers of test animals used ($N = 20$). Truncated curves (Fig. 3.2b) are usually explained in biological terms, viz. that the concentration is insufficient to kill the more resistant individuals of the distribution. In dispersant testing, such changes in slope are often indicative of losses of toxin occurring during the experiment through, for example, evaporation. Fig. 3.2c represents a response resulting from an apparent increase in the toxicity of a solution with time; this may occur when surfactants are degraded to more toxic components.

In an experiment involving several different concentrations, estimates of the median survival time are made for each of these concentrations, and by plotting these times against their respective concentrations it is possible to

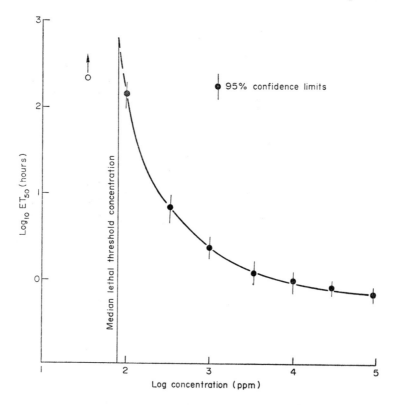

FIG. 3.3. A typical toxicity response curve for an oil dispersant.

establish a toxicity response curve (Fig. 3.3). The suitability of logarithmic scales for both time and concentration axes has been commented on by numerous workers, although the semi-logarithmic and reciprocal transformations have also been used. It is extremely important to establish the toxicity response curves, since fundamental differences between the shapes derived from two dispersants can indicate a difference in their mode of toxic action. Such differences have considerable bearing on the evaluation of the ecological risks arising from the use of the dispersants. In extreme cases, attempts to rank dissimilar dispersants are invalid.

From the toxicity response curve it is possible to derive one single measurement to describe the toxicity of a dispersant under particular test conditions, usually the concentration that has an effect (EC_{50}) or is lethal (LC_{50}) to 50% of the test organisms after a given period. The most commonly used are the 24-hour LC_{50}, 48-hour LC_{50} and 96-hour LC_{50}. These times are entirely arbitrary and, because they markedly influence the value of the LC_{50}, Sprague[19] recommends, as the most meaningful parameter with which to specify the effect of a poison, the median lethal threshold concentration. This is the concentration at which the toxicity ceases to be affected by further exposure.

FACTORS AFFECTING THE SHAPE OF THE TOXICITY RESPONSE CURVE

By definition, a standard bioassay technique must operate under certain biological and physical constraints. Before these conditions can be properly imposed, their significance in relation to the test result should be fully understood. Following the recommendations of Doudoroff *et al.*,[6] workers in fresh water have been urged to specify the size and number of the test species, the pH, hardness and temperature of the test solution, and numerous other pertinent measurements. These parameters are no less significant in the marine environment; some of the more important aspects are mentioned below.

Generally, few suitable *species* of marine organism are available to the investigator because the standard test species should be small, numerous, easily handled and readily available throughout the year. The major problem of supply has been the inability to culture a suitable marine species routinely, and animals must therefore be collected from the wild. The methods of capture and the treatment of animals after capture are critical to their well-being.[15] The availability of a species throughout the year is a requirement that is as important as it is difficult to meet. Owing to movements associated with spawning, feeding and changes in temperature and salinity preferences, species abundant at one season may be scarce or absent at another. Sessile species present different problems. For example, the bivalve molluscs *Mytilus* and *Ostrea* have been avoided for acute toxicity tests because it is diffcult to establish a simple criterion of effect for these species. Other species have been discarded because of changes in the size distribution of the population during the year, thus making it impossible always to use animals of a given size or size range.

The *condition* of the test animals is also important. Many species maintained in the laboratory suffer a decrease in viability even though they are feeding normally. Where food is offered it is important to ensure that all animals are accepting it; starved animals tend to be more susceptible to toxins (Fig. 3.4a). Similarly, unfed animals should be used quickly. Sex, developmental stage,[4,22] and the stage of the moulting cycle of crustaceans[20] have all been shown to influence the response of the animals to toxins.

Physical variables should be standardised where possible.

The *temperature* of the test solutions affects the rate of evaporation and degradation of oil and dispersants and the rate at which the animals die

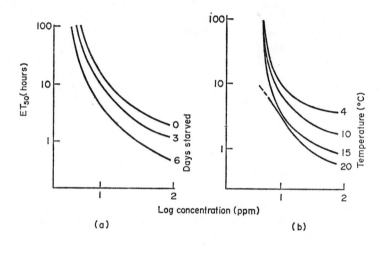

(a)   (b)

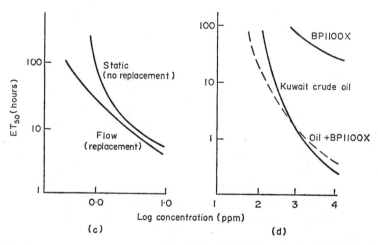

(c)   (d)

FIG. 3.4. The influence of biotic and abiotic factors on the toxicity response curve. (a) Effect of starvation on the toxicity of BP 1002 to plaice larvae. (b) Effect of temperature on the toxicity of BP 1002 to plaice larvae. (c) Effect of replacement on the toxicity of BP 1002 to *Crangon*. (d) Effect of oil/dispersant mixtures on *Crangon*.

(Fig. 3.4b). The effects of acclimatising the test species from seasonal ambient temperatures to a fixed test temperature are unknown, but the marked seasonal changes in susceptibility of some species[5,13] indicate the need for further investigation.

Salinity has a modifying effect on the toxicity of detergents[7] and it is likely to affect dispersants in a similar way.

Changes in the toxicity of dispersants in sea water due to losses from evaporation are well documented and similar losses from evaporation and degradation are known for crude oils, particularly in static systems where these substances are lost or inactivated. The use of a continuous-flow dosing apparatus overcomes these difficulties when the dispersant forms a stable emulsion in sea water (Fig. 3.4c). However, with more recent dispersant formulations, materials like crude oils tend to separate out from sea water as a discrete surface layer so that continuous and static dosing tests which do not provide vigorous agitation of the solutions underestimate their toxicity; the animals survive in the comparatively clean water below. Homogenisation of dispersant and crude oil mixtures can enhance the toxicity of both components (Fig. 3.4d).

COMPARING THE TOXICITIES OF DISPERSANTS

If the toxicity response curves of several dispersants tested under the same conditions are compared (Fig. 3.5), it is apparent that their relative toxicities are not constant with time, and it follows that rank orders established at different times differ significantly. The standard procedure adopted at Burnham-on-Crouch is to compare median lethal concentrations of dispersants at 48 hours (48-hour LC_{50}) and the median lethal threshold concentrations; the 48-hour LC_{50} is retained to allow comparisons with earlier determinations made at this laboratory.[16,17] It is significant that after 48 hours little change occurs in the rank order of most dispersants. Furthermore many of the effects of abiotic and biotic variables (see Fig. 3.4) are exerted on the lower portions of the toxicity response curve, with the threshold varying only slightly.

A consequence of maintaining a test for a long time is that the effective concentration of a dispersant is correspondingly lower. There is considerable justification for defining the toxicity of a dispersant for offshore species by means of the median lethal threshold concentration, because these low concentrations can result from the extensive dilution of the dispersant under the practical conditions of use at sea. This does not necessarily apply to littoral species, where exposure to very high concentrations of dispersants is probable. The toxicity response curves described earlier should not be used to assess the effects of dispersants on these species, since the time to death in continuous immersion does not bear any relation to the *effective* time to death which may result from a short exposure followed by immersion in clean sea water (Fig. 3.6). The acute toxicity of dispersants to littoral species is more properly determined using a dose and recovery technique.[13]

The suitability of any toxicity-testing technique can only be measured in terms of its standing in relation to other techniques and, ultimately, by how the results can be interpreted under field conditions.

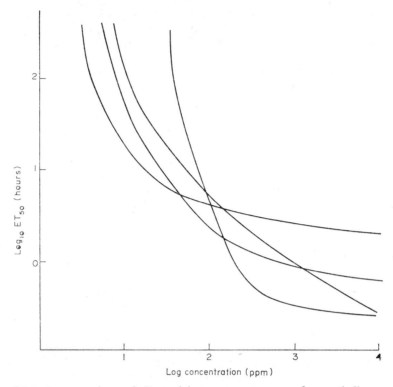

FIG. 3.5. A comparison of the toxicity response curves of several dispersants, using *Crangon crangon*.

At present, there are insufficient data from enough dispersants to allow comparisons of techniques. However, it is desirable that a standard technique should produce the same ranking order of a number of dispersants, irrespective of the test species. The toxicity data for ten dispersants given by Portmann and Connor[17] can be rearranged to produce a rank order with respect to their standard test species, the brown shrimp, *Crangon crangon* (Table I). This order can be compared with those given by three other species using the rank correlation coefficient r, where

$$r = 1 - \frac{6 \sum d^2}{n(n^2 - 1)} \quad \text{and} \quad \begin{aligned} n &= \text{number of ranks} \\ d &= \text{difference between any two rankings.} \end{aligned}$$

The rank order of all three species of crustacea showed close agreement, but significant differences existed between these and the bivalve mollusc, *Cardium edule*. However, the toxicities of these ten dispersants ranged over only one order of magnitude and should be compared with the toxicities of other dispersants which exhibit a range over four orders of magnitude (Fig. 3.7). If the rank orders of *Crangon* and *Cardium* towards dispersants covering this wide range of toxicities are compared, there is very good agreement (Table II).

However, La Roche et al.,[9] in defining a standard bioassay procedure for the evaluation of oil and oil dispersant toxicity, found a significant difference

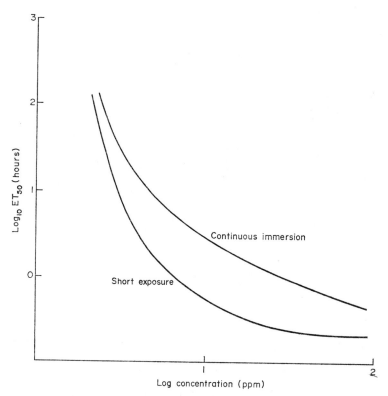

FIG. 3.6. A comparison of times to death of herring larvae resulting from a short exposure to, or continuous immersion in, BP 1002.

TABLE I

A Comparison of the Rank Orders of Ten Dispersants to Four Species of Marine Animals[17]

Dispersant	Crangon 48-hr LC_{50}	Rank order			
		Crangon	Pandalus	Carcinus	Cardium
Slickgone 2	(3·5)	1	1	4	4
BP 1002	(5·8)	2	3	1	8
Slickgone 1	(6·6)	3	2	6	5
Gamlen OSR	(8·8)	4	7	3	2
Essolvene	(9·6)	5	5	2	7
Polyclens	(15·7)	6	4	5	9
Cleanosol	(44·0)	7	8	7	3
Slix	(119·5)	8	6	8	1
Atlas 1909	(120·0)	9	9	9	6
Dermol	(156·0)	10	10	10	10
r		—	0·879	0·818	0·152

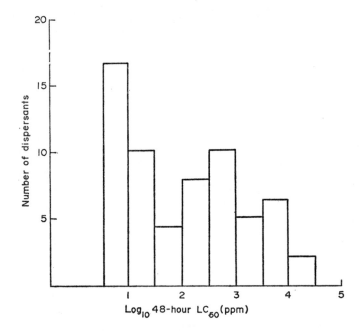

Fɪɢ. 3.7. The toxicity of several dispersants to *Crangon crangon* under standard conditions.

TABLE II

A Comparison of the Rank Orders of Ten Dispersants for
Crangon* and *Cardium

Dispersant	*48-hr LC_{50}*	*Crangon* *Rank*	*Cardium* *Rank*
Slickgone 2	(33–10)	1	3
BP 1002	(33–10)	2	5
Gamlen OSR	(33–10)	3	1
Cleanosol	(33–100)	4	2
Atlas 1901	(100–330)	5	4
Dermol	(100–330)	6	6
Polycomplex A	(100–330)	7	7
BP 1100	(1000–3300)	8	8
Corexit 7664	(3300–10 000)	9	9
BP 1100X	(> 10 000)	10	10

$$r = 0.843$$

in the rank orders of toxicity of the dispersants for the sandworm, *Nereis virens,* and the mummichog, *Fundulus heteroclitus,* even though the toxicities ranged over three orders of magnitude. There was no difference between the rank orders of toxicity of oil and of oil/dispersant mixtures for each species.

CONCLUSIONS

The basic principles for the design of tests for the measurement of toxicity in the aquatic environment have been defined repeatedly; if all the techniques which have been used followed them, then much of the confusion that exists in the presentation and interpretation of results would be avoided. Indeed, universal standardisation of a test much beyond these principles is not only impracticable but also meaningless, for the local conditions under which the dispersant is to be used should be an important consideration in establishing a technique. The detailed and accurate reporting of the methods and results of bioassays is as important for oil dispersants and oil as it is for other poisons. Even so, a true evaluation of the test is possible only if it can be referred to a standard dispersant or oil.

The aim of any ranking system is to express the result of a toxicity test by a single measurement and to use this value to grade the toxicity of a group of substances. From the shapes of the toxicity response curves it is evident that the concentration of a dispersant producing a lethal effect varies with many factors, of which time of exposure is perhaps the most important. For this reason, a test which determines the lethal threshold concentration and thereby eliminates the effect of time is the most useful and is one which, where possible, should be adopted. Where there is shown to be a constant relationship between the lethal threshold concentration and some other parameter, *e.g.* the 48-hour LC_{50}, the latter can be used as a more convenient tool.

The test adopted at Burnham-on-Crouch has concerned itself with comparing the toxicities of dispersants, and during the past five years has identified significant advances in the development of lower-toxicity dispersants. It must be stressed, however, that the results of this test can be used to evaluate the likely ecological impact of a new dispersant only with reference to local conditions.

REFERENCES

1. Alderdice, D. F., and Brett, J. R., Some effects of kraft mill effluent on young Pacific salmon, *J. Fish. Res. Bd Can.*, **14**, 783–95 (1957).
2. Bliss, C. I., The calculation of the time–mortality curve, *Ann. Appl. Biol.*, **24**, 815–52 (1937).
3. Boney, A. D., Experiments with some detergents and certain intertidal algae, *Field Studies*, **2** (suppl.) ('The Biological Effects of Oil Pollution on Littoral Communities', ed. J. D. Carthy and D. R. Arthur), 55–72 (1968).
4. Corner, E. D. S., Southward, A. J., and Southward, Eve C., Toxicity of oil-spill removers ('detergents') to marine life: an assessment using the intertidal barnacle *Elminius modestus*, *J. Mar. Biol. Assoc. U.K.*, **48**, 29–48 (1968).
5. Crapp, G. B., 'Laboratory Experiments with Emulsifiers', in *The Ecological Effects of Oil Pollution on Littoral Communities* (ed. E. B. Cowell), Institute of Petroleum, London, 1971, pp. 129–49.
6. Doudoroff, P., Anderson, B. G., Burdick, G. E., Galtsoff, P. S., Hart, W. B., Patrick, R., Strong, E. R., Surber, E. W., and Van Horn, W. M., Bio-assay methods for the evaluation of acute toxicity of industrial wastes to fish, *Sewage and Ind. Wastes*, **23**, 1380–97 (1951).
7. Eisler, R., Some effects of a synthetic detergent on estuarine fishes, *Trans. Am. Fish. Soc.*, **94**, 26–31 (1965).

8. Herbert, D. W. M., and Merkens, J. C., The toxicity of potassium cyanide to trout, *J. Exp. Biol.*, **29**, 632–49 (1952).
9. La Roche, G., Eisler, R., and Tarzwell, C. M., Bioassay procedures for oil and oil dispersant toxicity evaluation, *J. Wat. Pollut. Control Fed.*, **42**, 1982–9 (1970).
10. Litchfield, J. T., A method for rapid graphic solution of time–percent effect curves, *J. Pharmac. Exp. Ther.*, **97**, 399–408 (1949).
11. Nelson-Smith, A., The effects of oil pollution and emulsifier cleansing on shore life in south-west Britain, *J. Appl. Ecol.*, **5**, 97–107 (1968).
12. O'Sullivan, A. J., and Richardson, A. J., The *Torrey Canyon* disaster and intertidal marine life, *Nature, Lond.*, **214**, 448 & 541–2 (1967).
13. Perkins, E. J., The toxicity of oil emulsifiers to some inshore fauna, *Field Studies*, **2** (suppl.) ('The Biological Effects of Oil Pollution on Littoral Communities', ed. J. D. Carthy and D. R. Arthur), 81–90 (1968).
14. Perkins, E. J., Some effects of detergents in the marine environment, *Chem. Ind.*, 14–22 (1970).
15. Perkins, E. J., Some problems of marine toxicity studies, *Mar. Pollut. Bull.*, **3**, 13–14 (1972).
16. Portmann, J. E., 'The Toxicity of 120 Substances to Marine Organisms', Shellfish Information Leaflet No. 19, Fisheries Laboratory, Burnham-on-Crouch, Essex, UK, Sept. 1970 (mimeo).
17. Portmann, J. E., and Connor, P. M., The toxicity of several oil-spill removers to some species of fish and shellfish, *Mar. Biol.*, **1**, 322–9 (1968).
18. Smith, J. E. (ed.), '*Torrey Canyon*' Pollution and Marine Life, Cambridge Univ. Press, 1968, 196 pp.
19. Sprague, J. B., Measurement of pollutant toxicity to fish. I: Bioassay methods for acute toxicity, *Water Research*, **3**, 793–821 (1969).
20. Swedmark, M., Braaten, B., Emanuelsson, E., and Granmo, Å., Biological effects of surface active agents on marine animals, *Mar. Biol.*, **9**, 183–201 (1971).
21. Wilson, D. P., Long-term effects of low concentrations of an oil-spill remover ('detergent'): studies with the larvae of *Sabellaria spinulosa*, *J. Mar. Biol. Assoc. U.K.*, **48**, 177–82 (1968).
22. Wilson, K. W., in *Marine Pollution and Sea Life* (ed. M. Ruivo), FAO and Fishing News (Books) Ltd, London, 1972, 624 pp.

4

Toxicity Tests for Predicting the Ecological Effects of Oil and Emulsifier Pollution on Littoral Communities

J. M. BAKER

(*Oil Pollution Research Unit, Orielton Field Centre, Pembroke, Wales*)

and

G. B. CRAPP

(*Department of Zoology, University College, Cork, Ireland*)

INTRODUCTION

The purpose of toxicity testing is to predict the effects of a toxic substance on natural communities of animals and plants from the results of simple experiments. Tests are used to explore two areas of variability, that which is found in the relative toxicities and modes of action of different substances, and that which is found in the complexity of natural biological communities and their environment. An investigation will rarely be suitable for studying both these areas: thus, a test which enables us to rank different compounds in order of their toxicity quickly and accurately is not usually very useful for making ecological predictions, whilst ecological studies are generally unsuitable for making comparisons of relative toxicities. This is reflected in the recognition of an interface between the acquisition of laboratory data and the prediction of ecological effects.[11]

In this paper, two attempts to erode this interface are described, firstly in the case of oil pollution of saltmarshes, and secondly in the case of emulsifier pollution on rocky shores.

CASE 1: ASSESSMENT OF THE ECOLOGICAL EFFECTS OF OIL POLLUTION ON SALTMARSHES

In this case, a survey of actual oil spillage incidents was first carried out to identify the main problems involved. These emerged as follows:

1. A considerable variety of crude oils or oil products, which in turn vary with degree of weathering, may be stranded on saltmarsh vegetation.
2. According to their position relative to oil industry operations, marshes

23

may be oiled very rarely or may be affected by many small successive spillages or continuous low-level oiling from effluents.

3. The saltmarsh ecosystem includes different plant communities according to tidal level, and these communities in turn comprise many species, each with a seasonal cycle of growth, reproduction and death.

Thus, as pointed out in the introduction, two main areas of variability exist, the variability of the oil and the variability of the environment. A third problem can be separated in this case, that is, frequency of oiling.

Test for Relative Toxicities of Oils

30 × 40 cm turves of saltmarsh plants were kept in polythene-lined boxes (to prevent drying of the soil) in an unheated greenhouse. The relatively large size of turf was chosen to ensure an adequate central area free from the damage which inevitably occurs when the turf is cut. After an acclimatisation period of at least two weeks, turves were treated with different crude oils, oil products, oil fractions, or a crude oil at different stages of weathering. Turves were watered with fresh water regularly throughout the experiment, no attempt being made to follow the natural tidal cycle because of the very large practical problems. The actual species used, *e.g. Festuca rubra*, tolerates a wide range of salinities and may be found naturally growing in non-tidal communities.

Recovery from oil treatments was assessed by harvesting the turves two to five months after oil treatment, and obtaining a dry weight for healthy vegetation. Immediate post-treatment effects were misleading as in most cases the aerial shoots died, whatever the oil type. Recovery depended upon growth of new shoots from growing points protected from the oil by plant litter or soil, and in general was worst in the case of highly penetrating oils which affected the growing points.

Subsequent assessment of this type of test shows that it is very laborious, time- and space-consuming for ranking oils, and yields relatively little additional information of use in making ecological assessments. The data obtained, for example, on relative toxicities of crude oils did not differ significantly from that published by Ottway,[19] who ranked crude oils by an easier test using *Littorina littoralis*. In fact, as a general observation, relative toxicity information seems independent of the test organism involved, and the criteria for choosing experimental organisms should therefore be convenience and speed. For example, Nelson-Smith[17] used yeast in assessing emulsifiers.

Tests for (a) *Importance of oil spillage frequency*
          (b) *Importance of biological and environmental factors (including relative tolerance of species to oils, recovery rates, interspecific competition following oiling, and seasonal effects)*

These two aspects of the problem are not well suited to laboratory investigations. First, the experiments must take a long time, two years at least, if spillage frequency, recovery rates and seasonal effects are to be adequately covered. Second, it is doubtful whether experimental material can be kept under laboratory or greenhouse conditions for such a long time without degeneration. Third, it is very difficult to simulate continuously varying natural conditions for such a long time. Fourth, large numbers of species are involved, and these should be treated together in order to observe post-pollution interactions.

In short, it is completely impractical to attempt laboratory investigations of these factors, and therefore field experiments were used. Saltmarsh vegetation is particularly suitable for such experiments, for as the plants are fixed it is possible to treat a plot and observe the same individuals for months, without problems of migration. The same type of experiment might therefore also be suitable for fixed populations of animals such as barnacles.

The Field Experiments

An experiment using eight treated plots in each of three different saltmarsh communities was designed to give information on the factors mentioned above. A plan is shown in Fig. 4.1.

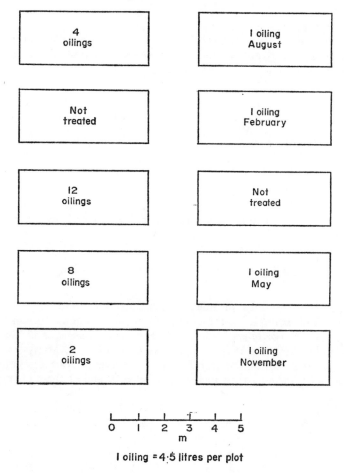

I oiling = 4·5 litres per plot

FIG. 4.1. Plan of an experiment using eight treated plots used in each of three different saltmarsh communities.

The oil used, Kuwait crude, was thought to be a suitable choice for field experiments because it is commonly encountered and its physical and chemical properties and toxicity, as shown by toxicity ranking tests, are all intermediate.

Some examples of information gained from this field experiment are given below:

(a) The relative tolerance of the many plant species forming a saltmarsh community:

Group 1 (very susceptible)
Shallow rooting, usually annual, plants with no underground storage organs, quickly killed by a single oil spillage, *e.g. Salicornia* spp.

Group 2 (susceptible)
Shrubby perennials with exposed branch ends which are badly damaged by oil, *e.g. Halimione portulacoides*.

Group 3 (susceptible)
Filamentous green algae. Though filaments are quickly killed, populations can recover rapidly by growth and vegetative reproduction of any unharmed fragments or spores, *e.g. Vaucheria* spp.

Group 4 (intermediate)
Perennials which usually recover from a spillage or up to four light experimental oilings, but decline rapidly if further oiled, *e.g. Puccinellia maritima*.

Group 5 (resistant)
Perennials which have a competitive advantage in vegetation recovering from oil, owing to fast growth rate and mat-forming habit, *e.g. Agrostis stolonifera*.

Group 6 (resistant)
Perennials, usually of rosette habit, with underground storage organs (*e.g.* tap roots). Most of them die down in winter, *e.g. Armeria maritima*.

Group 7 (very resistant)
Perennials of group 6 type which have in addition a resistance to oil at the cellular level and have survived 12 successive monthly oilings, *e.g. Oenanthe lachenalii*.

(b) The recovery rates of oiled plants and post-pollution changes in dominance. For example, in the upper saltmarsh community dominated by the rush *Juncus maritimus*, oiling first eliminates *Juncus*, and the grass *Agrostis stolonifera* then invades the mats of dead rush. Many successive spillages in turn reduce the *Agrostis*, and the highly resistant umbellifer *Oenanthe lachenalii* may then increase its cover. Some of these changes are shown in Fig. 4.2.

(c) The number of oil spillages that can be tolerated before the vegetation is all killed. For example, in the grazed *Puccinellia* zone, more than four monthly spillages results in persistent bare mud.

(d) The seasonal effects. Winter oiling can substantially reduce germination of seeds, and this is particularly noticeable in the case of annuals such as *Salicornia*. Spring oiling reduces the flowering of many species, and thus reduces seed production.

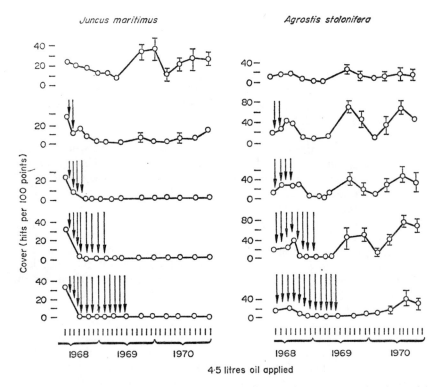

FIG. 4.2. Changes in cover of *Juncus maritimus* and *Agrostis stolonifera* with successive oilings.

Relationship between Different Types of Test

Data from initial toxicity-ranking tests are useful when deciding which pollutant to use in a field test, for, owing to the large size of field experiments, it is clearly not feasible to repeat them with many different oils. Relative toxicity data plus detailed knowledge of the effects in the field of one or more of the oils is a useful combination for making ecological predictions.

In some cases the greenhouse experiments using turves, which are really intermediate between the ideal simple toxicity test and the large field experiment, can be usefully used to follow up particular effects noticed in field experiments. For example, the phenomenon of growth stimulation, first noticed following Kuwait crude oil treatment in the field, was confirmed by greenhouse experiments under controlled conditions and the active compounds traced to the fractions boiling above 400°C. Similarly, field experiments smaller than that shown in Fig. 4.1 may be used to investigate particular problems or check predictions made about particular oils; for this purpose a simple treated plot may be sufficient.

Ecological Predictions

The results of the large field experiment were, whenever possible, compared with observations following oil spillages (a summary of these is given by

Baker [1,2,3]) and the two sets of data correspond well. Regular recording of experimental plots has now continued for nearly five years, so prediction of oil spillage effects within this time scale can be accurately made. For example, it can be predicted that recovery of saltmarsh vegetation from one oiling with Kuwait crude takes place within two years and that, before recovery is complete, reduced populations of annuals can be expected. Using relative toxicity data, it can also be predicted that recovery will be faster in the case of weathered crude oil,[5] and fresh crude oils ranked less toxic than Kuwait crude.

Longer-term predictions, which are relevant mainly in areas where successive spillages or effluent discharges occur, are more difficult. The experiment on successive spillages, and observations on saltmarshes near oil terminals or effluent outfalls, all indicate that chronic pollution causes a slow decline, but owing to the large variation possible with length and season of recovery periods between oilings, the rate is very variable.

From the tolerance grouping of saltmarsh plants given, it may be predicted that in cases of chronic pollution annual plants will disappear first and that rosette-type perennials with large underground systems will disappear last. This effect has been noticed on a small area of saltmarsh in Milford Haven. Further, it may be predicted that communities poor in number of species are particularly vulnerable to baring of the soil and subsequent erosion. For example, the *Spartina* community, being typically composed of only one species of flowering plant which is of intermediate tolerance, can be reduced to bare mud by eight successive spillages. A midmarsh community dominated by *Puccinellia* may lose all the *Puccinellia* after four successive spillages, and an upper marsh community dominated by *Juncus* may lose all the *Juncus* after only two spillages, but these two communities are nevertheless buffered by a variety of other more resistant species which can up to a point replace the lost dominants. Thus, bare mud appears later in these communities and there is less risk of erosion.

CASE 2: ASSESSMENT OF THE ECOLOGICAL EFFECTS OF EMULSIFIER POLLUTION ON ROCKY SHORES

The same basic problem was encountered in both this case and the saltmarsh study, but here it was expressed in these terms: laboratory tests may be used to compare the toxic effects of different compounds and the relative susceptibility of different species, but it is not possible to predict ecological effects from these results alone. The different ways in which the problem was defined and investigated reflected the great differences between the two kinds of habitats and communities; thus, the rocky shore study was seen as being mainly a zoological problem, whilst a botanical approach was needed for the saltmarsh. Similarly, field experiments could be readily carried out upon the rooted plants of the saltmarsh, but were much more difficult to design for the rocky shore, where the surface is usually broken and irregular, and where the dominant species range from attached plants and animals to large roaming snails.

Therefore a simple laboratory toxicity test was designed to determine the relative susceptibilities of different species to a single emulsifier, BP 1002.

Simultaneously, information was collected in the field after pollution incidents. It was not feasible to design a field experiment which was on a large enough scale to reproduce real spillage conditions and be little affected by migrations, and small enough to cover a fairly homogeneous and easily counted community, and so most of the field mortalities were assessed on accidentally polluted shores. In these cases a large area of shore was affected, but accurate pre-pollution data were not usually available and the intensity of emulsifier application was difficult to determine. Some small scale field experiments were made to obtain more accurate mortality figures.

In the account of this investigation which is given below, the design and the results of the laboratory tests are described first. An assessment is then made of the relationship between laboratory results and field mortalities, and cleaning incidents are assigned to three categories of varying severity. Some of the factors which can affect the accuracy of this correlation of laboratory and field results are discussed, followed by a description of the way in which laboratory tests were used to predict the ecological effects of a new emulsifier.

Design of the Laboratory Tests

Laboratory toxicity tests may be divided into two categories, those which use a constant time interval and various concentrations of the toxin, and those which use a constant concentration of the toxin for various times. In the former case the results may be expressed (in aquatic organisms) in terms of the LC_{50}, the concentration at which 50% of the organisms are killed, and in the latter case the TD_{50} is determined, the time interval which elapses until 50% of the organisms are dead. The LC_{50} test was chosen for the present work, for it is simpler to set up and more widely used in the case of littoral species. The determination of TD_{50} is most useful when the moment of an organism's death can be clearly observed, as happens with shrimps or larval forms, or when the toxin cannot be satisfactorily diluted, as happens with oil in sea water.

It was necessary to decide on a suitable time interval for the LC_{50} test, for in similar experiments this had varied from 30 seconds to 96 hours[20,24]. After some preliminary experiments a one hour interval was chosen, for this seemed to minimise the rather uncertain risks of the emulsifier separating out of the water, or of its very toxic fraction evaporating away.

The animals used in the experiments were chosen because they could be collected in quantity and easily handled and observed in the laboratory, whilst they were also common and important inhabitants of the rocky shores where the effects of pollution were studied in the field. The species investigated were the topshells *Monodonta lineata* and *Gibbula umbilicalis*, the periwinkles *Littorina littorea*, *L. littoralis* (*obtusata*) and *L. saxatilis rudis*, the dogwhelk *Nucella* (*Thais*) *lapillus*, the limpet *Patella vulgata*, the mussel *Mytilus edulis*, and the barnacles *Chthamalus stellatus*, *Balanus balanoides* and *Elminius modestus*.

The emulsifier BP 1002 was used in most experiments, for it was probably the most widely used product, it was fairly similar in toxicity to most other products,[20,24,25] and supplies were readily available through the generosity of the British Petroleum Company. The animals were exposed to various concentrations of BP 1002 in sea water for one hour, after which they were thoroughly rinsed in sea water and left in this to recover. All these species retracted into the shell on encountering even very dilute solutions of the emulsifier, and it was necessary to keep them for a five-day recovery period in order to determine which had recovered and which had died.

This experimental method was very similar to the 'extreme' exposures of Perkins,[20,21] and possibly reproduced the conditions found when emulsifiers were used in the recommended way in the field: that is, sprayed on to the shore undiluted and quickly washed away with a water jet.

Results of the Laboratory Tests

The results of toxicity tests with BP 1002 are shown in Fig. 4.3. For the sake of greater clarity in the following discussion, it is convenient to describe the susceptibility of each species in terms of five categories, namely:

1. Very resistant: the one-hour LC_{50} is greater than 5×10^5 ppm. Examples: *Monodonta lineata* and *Littorina littorea*.
2. Resistant: the one-hour LC_{50} lies between 5×10^4 and 5×10^5 ppm. Examples: *Littorina saxatilis rudis*, *Nucella lapillus*, *Gibbula umbilicalis* and *Chthamalus stellatus* in Fig. 4.3.
3. Moderately resistant: the one-hour LC_{50} lies between 5×10^3 and 5×10^4 ppm. Examples: *Littorina littoralis*, *Balanus balanoides* and *Elminius modestus* in Fig. 4.3.
4. Moderately susceptible: the one-hour LC_{50} lies between 5×10^2 and 5×10^3 ppm. Example: *Mytilus edulis*.
5. Susceptible: the one-hour LC_{50} lies between 5×10 and 5×10^2 ppm. Example: *Patella vulgata* in Fig. 4.3.

This order of resistance may be compared with the results of Perkins[20] and Smith[25] for many of these species. A detailed comparison is not possible because of variations in experimental technique, but in general terms the orders of resistance are similar. The discrepancies which can be found probably result from geographical and seasonal variations, and perhaps others such as the age and condition of the animals used.

The Meaning of the Laboratory Tests in Ecological Terms

The critical part of this study was the relating of laboratory results to observations made of the effects of pollution in the field, and this proved to be the most difficult part to carry out satisfactorily. Field observations were based mainly on Cornish shores polluted by *Torrey Canyon* oil, described by Smith;[25] the shores at Llanreath and Hazelbeach polluted by oil from the *Chryssi P. Goulandris*, described by Nelson-Smith;[15,16] the shore at Hazelbeach polluted by oil from a refinery accident, described by Crapp;[9] and an experimental pollution at Greenala Point in Pembrokeshire, described by Crapp.[7] It was not possible to tabulate the results in a simple form, because various factors complicated their interpretation. For instance, different methods were used to record mortalities; notes were made on the *Torrey Canyon* shores, a semi-quantitative abundance scale was used at Llanreath and Hazelbeach, and quadrat counting at Greenala Point. A very important point was that animals may not be killed by emulsifiers, but affected in other ways. The littoral snails—topshells, periwinkles and dogwhelks—all retracted into the shell on

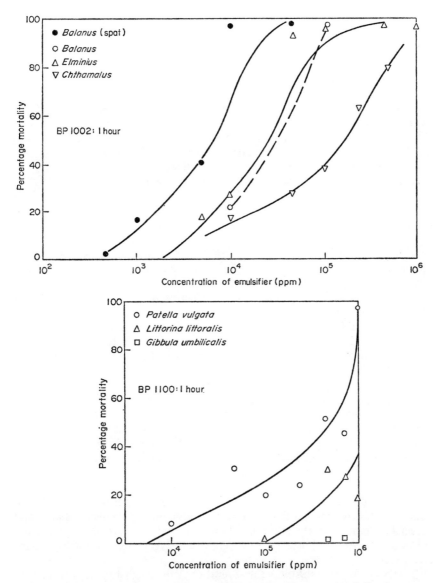

FIG. 4.3. The percentage mortalities recorded in various species following one-hour exposure to BP 1002 and BP 1100.

encountering emulsifier, and were dislodged from the rock surface, whilst other species remained attached. Thus, the numbers of these snails found after the spillage depended not only on how many survived poisoning, but also upon how many were able to return to the shore after a period of being washed and rolled helplessly about. This behaviour, an initial disappearance followed by the return of part at least of the population, has been observed after several cleaning incidents.[4,9,25] Factors other than the intensity of emulsifier pollution would affect mortalities in other ways as well, for animals might be protected from the severest exposure by living in a sheltered habitat, or they might be smashed by water from high-pressure hoses.

Despite these difficulties, enough information was collected on polluted shores to show that the relative susceptibilities determined in the laboratory were reflected in the field mortalities. For instance, on the *Torrey Canyon* shores *Patella vulgata* was killed in very large numbers, while *Monodonta*, *Nucella*, *Gibbula*, *Littorina* and *Chthamalus* were much more resistant, with *Mytilus* in an intermediate position.[25] At Hazelbeach the extreme sensitivity of the limpets was clearly contrasted with the initial disappearance and subsequent recovery of the snails.[9] The order of decreasing resistance *Chthamalus–Balanus–Mytilus–Patella* was found in the Greenala Point experiment.[7]

The experience gained in the study of these and other shore cleanings made it possible to define three categories of cleaning intensity and its consequences as follows:

1. When emulsifiers were used sparingly and were washed away with large quantities of water, even the 'susceptible' limpets were not severely affected. Numbers of other species were sometimes depleted, but the grazing population remained sufficiently numerous and active to prevent invasion of the shore by green and brown algae (see below). Recovery took place fairly rapidly, probably within two or three years, and ecological relationships on the shore were not greatly disturbed. This happened on the shores at Llanreath and Hazelbeach which were cleaned after the *Chryssi P. Goulandris* spillage.[15,16]

2. Increased use of emulsifiers resulted in very heavy mortalities in *Patella vulgata*, a species 'susceptible' to emulsifier poisoning. Other species suffered some mortality, particularly if 'moderately susceptible' or 'moderately resistant', and littoral snails retracted into their shells and were dislodged from the shore. These molluscs returned to the shore if they survived poisoning and a period of inactivity and were not carried too far away. Green and brown algae invaded the shore, but this development was restricted by the grazing activities of any surviving limpets, together with other herbivorous molluscs such as periwinkles and topshells. This happened at Hazelbeach following the refinery accident pollution, and on experimental squares at Greenala Point.[7,9] In the latter case limpet mortalities were almost total and no other grazing molluscs were present, and a very heavy growth of algae developed.

3. Heavy use of emulsifiers for several days in succession resulted in very heavy mortalities among not only the 'susceptible' *Patella vulgata*, but also the 'moderately susceptible' *Mytilus edulis* and the 'moderately resistant' *Balanus balanoides*. Periwinkles, topshells and dogwhelks were dislodged and many were probably killed by being exposed to emulsifier. In the more extreme cases heavy mortalities occurred among the 'resistant' *Chthamalus stellatus*. Many shores polluted by *Torrey Canyon* oil were cleaned to this extent, notably Trevone and Porthleven,[25] and the shore was subsequently invaded by green and brown algae.

This classification gave some idea of what the results of laboratory toxicity tests meant in ecological terms, and it was possible to use laboratory tests of a new emulsifier to make predictions about the ecological effects that it would have. However, before this is described some of the variables affecting the laboratory–field relationship, and the accuracy of predictions made from it, will be considered. These may be divided into the short term factors operating within the time-scale of the laboratory tests, and the long term factors which only become apparent over longer periods.

Factors Affecting the Accuracy of Short-term Predictions

Firstly, emulsifiers are used to remove spilt oil from the shore, and mortalities may be affected by the toxicity of oil alone, or the influence of oil when mixed with emulsifier. Oil alone was found to be relatively harmless on *Torrey Canyon* beaches,[25] and this has been confirmed, except in the case of light and volatile oils such as gasoline and kerosene, by field and laboratory experiments.[7,10] It is not clear at present whether the presence of oil increases or decreases the toxicity of emulsifier solutions, and probably either may happen with different materials and in different conditions.[18] However, it is probable that the presence of oil has a relatively small effect compared with the degree of toxicity found in the emulsifier itself, and in most beach-cleaning operations a considerable proportion of the emulsifier solution has little oil mixed with it.

Secondly, Perkins[20] drew attention to the possibility that seasonal variations in susceptibility may affect the results of toxicity tests, and this was investigated by Crapp[8] using the methods and the species (except the barnacles) described above. The greatest variations were found in *Monodonta lineata*, *Gibbula umbilicalis*, *Littorina littoralis* and *Nucella lapillus*, as illustrated in Fig. 4.4. These variations are only sufficient to transfer a species from one 'resistance category' to the next, and will be most marked in influence when two species vary in an opposing fashion: for instance, in summer *Monodonta lineata* was more resistant to BP 1002 than *Nucella lapillus*, and in winter the reverse was true. Such seasonal variations will affect the results of toxicity tests when similar species or materials are investigated, but should only be a minor source of inaccuracies on a larger scale (such as a comparison between *Monodonta*, *Mytilus* and *Patella*, or between BP 1002 and BP 1100, as described below). It was not possible to explain these seasonal variations in simple physiological terms.

Thirdly, care must be taken to observe whether an animal's behaviour is such that factors other than emulsifier toxicity will affect its survival in the field. An obvious example of this is the behaviour of the littoral snails referred to earlier. These retract into the shell on encountering even low concentrations of emulsifier, the activity of *Monodonta lineata*, for example, being inhibited by only 10–16 ppm of BP 1002.[8,25] The parameters of inactivation can be defined, for Perkins[20,21] has determined the thresholds of inactivation in several snails, and Crapp[8] found that the rate of recovery was characteristic for each species and could be statistically defined in terms of its relationship with mortality, but other factors will affect field mortalities. The retracted snail will roll or be swept away from the shore, and the numbers that recover will depend upon the intensity and duration of emulsifier pollution. The animals will be unable to avoid predators whilst inactive, and the abundance and activity of these will be only loosely linked with the intensity of pollution.

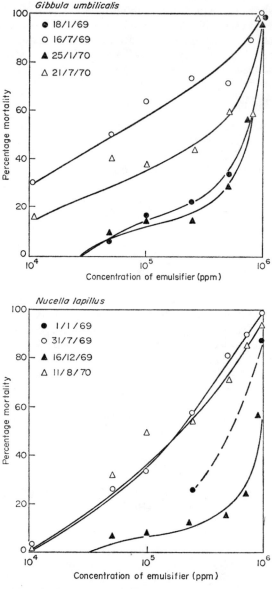

FIG. 4.4. Seasonal variation in the response of four common littoral species to exposure to BP 1002 for one hour at varying concentrations.

The animals may be carried by water movements into habitats where they cannot recover or return to the shore, and this will be unrelated to pollution. This kind of behaviour can obviously upset ecological predictions, and it was fortunate that in the studies described earlier it was the return of these hardy species to the shore that was significant, rather than their disappearance.

Differences in the way the tests are set up can affect the results, and several

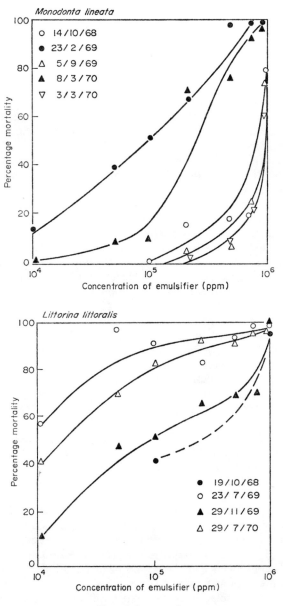

FIG. 4.4—*contd.*

aspects of this problem have been discussed by Perkins.[22] One example relevant to the present work is the fact that when *Mytilus edulis* is collected, the byssal threads must be cut and not torn: this precaution was neglected in the study of seasonal variations in susceptibility and this might account for the rather erratic results obtained.[8] In the experiments described above, slightly higher mortalities were recorded when emulsifier solutions were poured on to active and crawling snails rather than on to retracted ones, and the former arrangement was used as the standard method.

Factors Affecting the Accuracy of Long-term Predictions

Obviously the short-term toxicity test can only be used to predict short-term field mortalities. Will long-term effects influence the accuracy of ecological predictions based on short-term laboratory results? Three kinds of long-term effect have been identified, the delayed mortalities and the sublethal effects caused by emulsifiers, and the ecological changes which follow the damaging of part of the littoral community.

Delayed mortalities following emulsifier application were first observed by Wilson[29] in the larvae of *Sabellaria spinulosa:* animals which recovered within a few days, and were apparently healthy, died some weeks later. Perkins[21] found similar behaviour in *Littorina saxatilis, L. littorea* and *Nucella lapillus*, and concluded from his results that short-term toxicity tests gave a poor indication of long-term mortalities. However, this is not so important as might at first appear, for two reasons. Firstly, the delayed mortalities are probably only a small proportion of the total deaths. There is not a great deal of evidence on this point, but after the Hazelbeach cleaning referred to earlier Crapp[9] found that the limpet *Patella vulgata* was initially reduced in numbers from more than 100 per m² to around 21–25 per m², and this then declined to about 5 per m² over the next eight months. Some other species appear to have behaved in a similar way, but the results for these are only semi-quantitative. Secondly, the ratio of long-term to short-term mortalities probably does not vary to a great extent from one species to another. Thus delayed mortalities may well vitiate predictions made from absolute mortalities recorded in the laboratory, but they should be less important when laboratory mortalities are defined in purely relative terms and compared with field events over an adequate time scale.

Animals which survive emulsifier poisoning may suffer an impaired vitality, reflected in reduced rates of feeding and growth. This can be shown in experimental work: for instance, Perkins[21] found depressed growth rates in *Littorina saxatilis, L. littorea* and *Nucella lapillus* which had been exposed to emulsifiers, and Crapp[6,8] found a reduced rate of feeding in *Nucella* after exposure to BP 1002. However, little is known about how serious such sublethal changes are in natural communities. Bryan[4] studied the growth of *Nucella* after the *Torrey Canyon* spillage, and found that even on the severely polluted Porthleven reef the reduction in growth was no more than would be expected as a consequence of the period of inactivation. It seems likely that in the field observations which were described above any sublethal changes were too slight to be noted or to have much influence on the course of events.

Much more serious were the changes in community structure that were the result of the observed mortalities. The activities of each species affect the biology of others, and if a proportion of the shore population is removed, then the conditions affecting settlement, growth and mortality will change. The most striking example of this was the appearance of an algal canopy on shores where the herbivorous limpets had been killed. This was studied some years ago in a field experiment at Port Erin, described by Jones,[12] Lodge,[13] Southward[26] and others, in which the limpets were smashed with hammers.

A dense growth of green and brown algae soon appeared, and many species, notably the barnacles, were greatly reduced in abundance under the weeds. Limpet spat began to settle on the rocks, and as their numbers increased fewer algal sporelings escaped grazing. After around five years the limpet population had re-established itself to the point where few algae were able to settle and grow, and the appearance of the shore returned to normal as the older weeds died.

Emulsifiers such as BP 1002 appear to be almost as effective as hammers for destroying limpets, and after all but the lightest cleaning operations an algal invasion appeared. For example, Hazelbeach was cleaned in November 1968 and most of the limpets were killed, and a large number of green *Enteromorpha* plants were found in the following March. These green algae had been replaced largely by the brown *Fucus vesiculosus* in September. A similar sequence of events was observed in the Greenala Point experiment, and in both cases very large numbers of young limpets were found under the algal canopy, which probably indicates that conditions for settlement and growth of the spat were much more favourable than on the dry, barnacled rocks found before the incident.

Particular attention is given here to shores dominated by limpets, for it happened that most of those studied were dependent on limpet grazing for preventing algal invasion. In these cases it was found that there was little difference between the ecological consequences of the second and third categories of cleaning intensity which were described above. In both cases an algal canopy developed, and under this many species which survived second-category cleaning were greatly reduced in abundance: thus, after two or three years the two kinds of shores appear to be very similar and will recover their normal community in a similar way. This is an excellent illustration of the general principle that if a key species is particularly susceptible to pollution, then even a moderate level of pollution can have the most drastic effects.

The spores and larvae that settled and grew on polluted shores once cleaning was over were probably almost entirely derived from other, unpolluted shores, and the effects of emulsifier poisoning studied here were doubtless minimised by the fact that pollution was not continuous, and only affected a part of the littoral ecosystem. The affected shore was largely recolonised by the planktonic offspring of the plants and animals of more distant areas, and the nature of the recolonisation must depend primarily on the suitability of the shore for settlement and survival in each species, rather than on how many young the survivors can produce. Species with no planktonic dispersal phase in the life cycle will form an exception to this rule, and will be particularly vulnerable in terms of their ability to recover from pollution incidents.

The Use of the Toxicity Test for Ecological Predictions

Once a correlation between the laboratory and field toxicities of BP 1002 had been made, and taking into account the various difficulties described above, it was possible to test a new emulsifier, BP 1100, in the laboratory and make ecological predictions from the results. Three species were tested with this product, and once its low toxicity had been revealed by one-hour toxicity tests, ten-hour exposures were carried out as well. The results, illustrated in Fig. 4.3, indicated that *Patella vulgata* was slightly more resistant to BP 1100 than *Gibbula umbilicalis* was to BP 1002, whilst *G. umbilicalis* and *Littorina littoralis* were very resistant to BP 1100. Therefore, assuming that both emulsifiers affect different species in the same order of susceptibility, species which are 'susceptible' to BP 1002 will be 'resistant' to BP 1100.

This can be defined in ecological terms by referring to the categories of cleaning intensity defined earlier. When BP 1002 was used, *Patella* was killed by cleaning in the second and third categories, and *Gibbula* only by cleaning in the third category. It appears, therefore, that if BP 1100 is substituted for BP 1002, then 'third category' cleaning will give 'second category' effects, and so on. Even if BP 1100 was used very heavily, it would be unlikely to kill many of the more resistant species, but it could kill most of the sensitive limpets, and it was emphasised earlier that on many shores this can lead to ecological consequences as severe as those which follow almost total destruction of the shore community.

The laboratory toxicity test can also be used to predict the ecological consequences of emulsifier pollution on hitherto untested species, provided that the susceptibility of these can be compared with that of species in which field mortalities have been investigated. Now that 'low toxicity' emulsifiers are being developed it is desirable that the susceptibility of many delicate species, such as hydroids or small crustaceans, should be determined as well as that of the 'littoral toughs' investigated in the present work.

The basic principle of this procedure is that no laboratory, or small scale field, toxicity test can be expected to reproduce all the conditions which apply in a real polluted ecosystem, and in the case of oils and emulsifiers real pollution incidents are frequently available for observation. The purpose of toxicity testing is therefore analysis, not reproduction, of real incidents, and the LC_{50} test is useful because it is simple and easy to replicate. The correlation of relative LC_{50}'s with field mortalities and with changes in community structure has been expressed here in very general terms because a number of variables affect it, summarised in the following list.

1. Differences in the way emulsifier is applied to the shore, such as spraying or pouring, whether hoses are used, whether fresh water is used, the state of the tide when cleaning is carried out.
2. The effects of oil on emulsifier toxicity.
3. Seasonal changes.
4. Behavioural changes (such as the inactivation of littoral snails).
5. Differences in laboratory test materials and procedure.
6. The influence of delayed mortalities in the field.
7. The influence of sublethal effects in the field.
8. The changes in ecological conditions which follow the destruction of part of the biological community.

It appears, from the work described here, that the last of these is much more important ecologically than the preceding seven. The rocky shore cannot, after an incident of acute pollution, be regarded as an isolated ecosystem, because there is a continuing pressure of settlement by the spores and larvae of many species. The climax community perpetuates itself through mechanisms which regulate settlement, growth and survival, and the extent of post-pollution disruption depends upon the amount of damage done to these mechanisms.

The investigation described here was primarily directed at defining the overall consequences of emulsifier pollution and relating these to a simple toxicity test. This has only been possible in very general terms because the relationship is affected by so many variables, and a great deal of further work could be carried out on the influence of these.

CONCLUSIONS

In the introduction the tests described here were referred to as an attempt to erode the interface between laboratory and field studies: how far do they succeed in doing this? Quite clearly, the tests described are poor bioassay techniques. The methods used for saltmarshes require turves or a large area of marsh and a long period of time, and whilst the laboratory testing of rocky shore animals takes less time and space, it is still very slow and cumbersome. Both tests use material from heterogeneous biological communities unless carried out on a small scale, and in both cases the results may be affected by seasonal changes.

The tests were devised by a process of working from ecological investigations in the field towards a simple bioassay technique and, as suggested in the introduction, it was unlikely that we would get all the way in one simple step. We believe that the kind of methods described here are most usefully regarded as a means of linking the results of bioassays with the results of ecological investigations. Toxicity testing can thus be regarded as a three-level procedure. Simple standardised bioassays, such as those described by Tarzwell,[27,28] can be used to rank the toxicities of a large number of materials and their constituents. A smaller number of the more interesting materials can then be tested by the methods outlined here, methods for making ecological assessments. The accuracy of the ecological predictions can be tested by using a very small number of materials in full-scale field investigations.

Finally, four critical properties of toxicity tests for making ecological assessments may be defined, and contrasted with the properties of the bioassay test for ranking toxic materials:

1. Each test must be designed for making predictions about a particular kind of community or habitat, and should be based upon field studies in that community or habitat.
2. The test must be capable of predicting mortalities among the key species of the community.
3. The effects upon mortalities of such factors as seasonal or behavioural variations must be appreciated.
4. The ecological consequences of these mortalities must be understood.

REFERENCES

1. Baker, J. M., 'The Effects of a Single Oil Spillage', in *The Ecological Effects of Oil Pollution on Littoral Communities* (ed. E. B. Cowell), Institute of Petroleum, London, 1971, pp. 16–20.

2. Baker, J. M., 'Successive Spillages', ibid., pp. 21–32.
3. Baker, J. M., 'Refinery Effluent', ibid., pp. 33–43.
4. Bryan, G. W., The effects of oil-spill removers (detergents) on the gastropod *Nucella lapillus* on a rocky shore and in the laboratory, *J. Mar. Biol. Ass. U.K.*, **49**, 1067–92 (1969).
5. Cowell, E. B., The effects of oil pollution on salt marshes in Pembrokeshire and Cornwall, *J. Appl. Ecol.*, **6**, 133–42 (1969).
6. Crapp, G. B., 'The Biological Effects of Marine Oil Pollution and Shore Cleansing', Ph.D. thesis, University of Wales, 1970.
7. Crapp, G. B., 'Field Experiments with Oil and Emulsifiers', in *The Ecological Effects of Oil Pollution on Littoral Communities* (ed. E. B. Cowell), Institute of Petroleum, London, 1971, pp. 114–28.
8. Crapp, G. B., 'Laboratory Experiments with Emulsifiers', ibid., pp. 129–49.
9. Crapp, G. B., 'The Biological Consequences of Emulsifier Cleansing', ibid., pp. 150–68.
10. Crapp, G. B., 'The Ecological Effects of Stranded Oil', ibid., pp. 181–6.
11. Edwards, R. W., Future research needs, *Proc. Roy. Soc. Lond.*, Ser. B, **177**, 463–8 (1971).
12. Jones, N. S., Observations and experiments on the biology of *Patella vulgata* at Port St. Mary, Isle of Man, *Proc. L'pool Biol. Soc.*, **56**, 66–77 (1948).
13. Lodge, S. M., Algal growth in the absence of *Patella* on an experimental strip of foreshore, Port St. Mary, Isle of Man, *Proc. L'pool Biol. Soc.*, **56**, 78–83 (1948).
14. Logan, J. W. M., and Perkins, E. J., Toxicity of Essolvene, *Marine Pollution Bull.*, **3**, 155–7 (1972).
15. Nelson-Smith, A., Biological consequences of oil pollution and shore cleansing, *Fld. Stud.*, **2** (suppl.), 73–80 (1968).
16. Nelson-Smith, A., The effects of oil pollution and emulsifier cleansing on marine life in south-west Britain, *J. Appl. Ecol.*, **5**, 97–107 (1968).
17. Nelson-Smith, A., 'Micro-respirometry and Emulsifier Toxicity', Field Studies Council Oil Pollution Research Unit, Ann. Rep., 1969.
18. Nelson-Smith, A., The problem of oil pollution of the sea, *Adv. Mar. Biol.*, **8** (1970),
19. Ottway, S., 'Some Effects of Oil Pollution on the Life of Rocky Shores', M.Sc. thesis, University of Wales, 1972.
20. Perkins, E. J., The toxicity of oil emulsifiers to some inshore fauna, *Fld. Stud.*, **2** (suppl.), 81–90 (1968).
21. Perkins, E. J., Some effects of detergents in the marine environment, *Chem. Ind.*, 14–22 (1970).
22. Perkins, E. J., Some problems of marine toxicity studies, *Marine Pollution Bull.*, **3**, 13–14 (1972).
23. Portmann, J. E., and Wilson, K. W., 'The Toxicity of 140 Substances to the Brown Shrimp and Other Marine Animals', Shellfish Information Leaflet, Ministry of Agriculture, Fisheries and Food, No. 22 (1971).
24. Simpson, A. C., Oil, emulsifiers, and commercial shellfish, *Fld. Stud.*, **2** (suppl.), 91–8 (1968).
25. Smith, J. E. (ed.), '*Torrey Canyon*' *Pollution and Marine Life*, Cambridge Univ. Press, 1968, pp. 196.
26. Southward, A. J., Limpet grazing and the control of vegetation on rocky shores, *B.E.S. Symposium*, **4**, 273–5 (1964).
27. Tarzwell, C. M., 'Toxicity of Oil and Oil–Dispersant Mixtures to Aquatic Life', in *Water Pollution by Oil* (ed. P. Hepple), Institute of Petroleum, London, 1971, pp. 263–72.
28. Tarzwell, C. M., Bioassays to determine allowable waste concentrations in the aquatic environment, *Proc. Roy. Soc. Lond.*, Ser. B, **177**, 279–85 (1971).
29. Wilson, D. P., Long-term effects of low concentrations of an oil-spill remover (detergent): studies with the larvae of *Sabellaria spinulosa*, *J. Mar. Biol. Ass. U.K.*, **48**, 177–82 (1968).

5

Toxicity Testing at Kristineberg Zoological Station

MARTHA SWEDMARK

(*Vestenskapsakadamien, Kristineberg Zoological Station, Sweden*)

Bioassay work with marine animals as test species has been performed at the Kristineberg Zoological Station for several years. Kristineberg Zoological Station is situated at the Gullmar Fjord on the west coast of Sweden. The studies are focused on the effects of surface-active agents and a part of this broader project has been the toxicity testing of oil dispersants and of mixtures of oils and dispersants. The work is financially supported by the Swedish Environment Protection Board and the Royal Swedish Academy of Sciences to which the station belongs.

The purposes of the toxicity testing done at the station are (1) to determine the *relative toxicities of the different materials* in standard form required by industry and government bodies, and (2) to provide *predictions of the ecological consequences of pollution* in marine environments.

The investigations are undertaken as *comparative* studies of toxic materials on a wide spectrum of marine animals, represented by fish, crustaceans and bivalves. The studies involve both adult animals and developmental stages. This is of great importance as the resistance of animals varies considerably during their life-cycle, the early phases generally being the most sensitive.

The *acute* or *lethal* toxicity is determined in *short-term* tests (96 hours). During the tests not merely mortality and survival times are recorded but also continuous observations of effects on various biological functions which are important for the survival of the animals in their natural environment. It is from such observations that conclusions of the ecological consequences of pollution on the animal communities can be made.

Chronic effects on the same biological functions are studied in *long-term* tests, running for several months in low concentrations corresponding to those which may occur in coastal waters.

A physiological approach is also attempted by the study of the action of surface-active agents on respiration, osmoregulation and accumulation in tissues and organs.

METHODS

The biotest technique used at Kristineberg is based on the *continuous-flow principle*. Continuous-flow aquaria systems are generally considered to

41

facilitate the maintenance of more stable conditions in the physico-chemical environment of the test tanks, *i.e.* of such factors as temperature, salinity, pH, oxygen content, NH_3 content, bacterial level, etc. Tests in continuous-flow aquaria should thus give more accurate results.

The use of running sea water is possible at Kristineberg because the station has a constant supply of *clean* sea water being pumped into the laboratory from a 40 m depth of the Gullmar Fjord. At this depth the salinity is quite constant, between 32 and 34‰, and the oxygen content is high, 80–90% saturation. The incoming sea water has a temperature variation throughout the year from 4° to 16°C but with no rapid fluctuations.

The *biotest equipment* (Fig. 5.1) is principally designed as described by Swedmark *et al.*[5] It consists of three levels of aquaria with a varying number (3–12) of test tanks at each level with a common dosing equipment and one control section. Each vertical section is alimented from its respective funnel where the mixing of standard solution and sea water takes place. To increase the mixing efficiency the funnels are equipped with glass beads on coarse nylon mesh.

In order to keep the concentrations of the tested material constant, the standard solution is added to the test aquarium by means of precision dosing

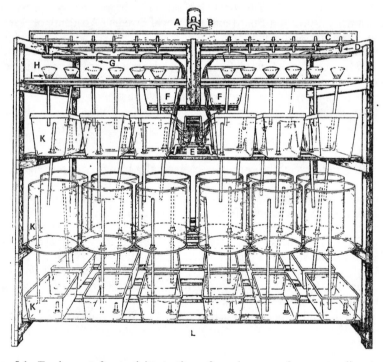

FIG. 5.1. Equipment for toxicity testing of marine organisms according to the continuous-flow principle. Sea-water inlet A, sea-water overflow outlet B, sea-water dosing apparatus C with dosing nozzles D, dosing pump E, standard solution tank F, standard solution dosing tubes G, mixing funnels H with glass beads I, test tanks K, sea-water outlets L.

pumps (electric piston pumps or microperistaltic pumps). The running sea water is distributed by siphons, valves or by a special dosing apparatus.[1] All tubing is of Tygon or Silicone.

Rectangular or cylindrical Perspex aquaria are used. Their volume varies between 10 and 60 litres, according to the size of the test animals. Cylindrical vessels are always used for testing eggs and larvae. Generally the number of adult animals in each tank is 10, but sometimes for practical reasons it is less.

The equipment here described permits simultaneous testing of several species in a number of different concentrations at uniform milieu conditions and yields a great quantity of data within the test time. It has a simple design and has proved to be of good reliability. It has not been considered necessary to incorporate safety devices.

The *standard solutions* of soluble material (surfactants, some oil dispersants) are made by dilution with distilled water. While testing non-soluble dispersants and crude oils the concentrated, undiluted products are used. These latter materials were very destructive to the tubing, which gave rise to some problems.

The emulsions (mixtures) of different oils and oil dispersants are made up by means of a stirrer running at 200 rev/min in the standard solution tank. For light fractions of refined oil, such as marine diesel oil, the proportions used between dispersant and oil were kept at 1:9 to give a tight homogeneous emulsion. For heavier fractions such as fuel oil or for crude oil, the corresponding proportions were changed to 1:1 to give a similar emulsion.

By the mixing methods used, surfactants, soluble and non-soluble dispersants as well as oil emulsions are well mixed with sea water, and the dispersions formed generally keep their homogeneity during their passage through the aquaria. With the crude oil, however, there is only poor mixing with sea water and a tight dispersion is not obtained. At the testing the oil was left to float on the surface of the water or to appear as droplets in a coarse dispersion. This was considered to correspond quite well to conditions in waters of low turbulence and without tides, and no other mixing method was tried.

Concentrations from 1 to 1000 ppm (mg/litre) are used as standard concentrations. As a rule each product is tested in duplicate series run in parallel. So far only one product has been tested at a time.

The *sea-water flow* is generally kept at 0·5 litre/min and there are only small variations in the flow (less than 1%). Regular control of the oxygen content shows that this flow is enough to give satisfactory oxygenation (more than 70%) at all levels.

As a rule, short-term testing is run at temperatures corresponding to normal sea-water temperature of the season, as the animals are then exposed to less extra stress. The temperatures used are 6°C, 11°C and 16°C. When necessary the temperature of the test tanks is kept approximately constant with the aid of an immersion heater or cooler equipment.

As the susceptibility of the animals has been shown to vary with temperature and season, we have found it preferable to make comparative short-term tests during the same season.

Short-term exposure always lasts 96 hours and is followed by a recovery period of at least 48 hours in clean sea water. The recovery period is particularly important when testing bivalves as they often manifest a delayed mortality.

Temperature, salinity and dosage rate are checked daily. The oxygen content is checked once per short-term test, or once a week in long-term tests. Mortality and effects on biological functions are recorded several times a day during short-term testing.

PRODUCTS TESTED

The most common surfactants used in the manufacturing of household and heavy-duty detergents have been tested, as also has a series of commercial oil dispersants. Both the older, highly toxic dispersants and lower-toxic products are represented.

Earlier products are represented by Fina-Sol SC, non-soluble, made by Fina SA, Belgium. Newer types are represented by the following products: Berol TL–188 and Berol TL–198, both soluble, manufactured by Modokemi AB, Sweden; BP 1100 and BP 1100X, both non-soluble, made by BP Trading Ltd, Great Britain; Corexit 7664, soluble, and Corexit 8666, non-soluble, both manufactured by Esso Chemicals, USA; Fina-Sol OSR–2, non-soluble, made by Fina SA, Belgium; and Polyclens Industrial TS7, non-soluble, manufactured by Polycell Products Ltd, Netherlands.

The following oil emulsions are tested: Corexit 8666 + light marine diesel oil; Corexit 7664 + light marine diesel oil; Corexit 7664 + fuel oil 4 (Swedish standard), low sulphur content; Corexit 7664 + Oman crude oil.

Oman crude oil has also been tested alone, only mechanically dispersed. This crude resembles the Kuwait crude in its properties except for having a low sulphur content.

SELECTION OF TEST ANIMALS

A variety of marine fish, bivalves and decapod crustaceans are used as test organisms. They all live in the littoral zone (0–40 m) and are caught in the Gullmar Fjord and nearby waters.

They have been selected according to the following criteria:

1. They should represent different taxonomic groups and also different modes of life and activity with regard to locomotion (mobile, sedentary) and nutrition (carnivorous, filter-feeders).
2. They should include species representing different trophic levels.
3. They should include species of economic importance (cod, lobster, mussels).
4. They should be susceptible and the biological functions studied should be easy to define and observe.

The choice may be limited for practical reasons such as availability in the region as well as ease of collection and adaptability to laboratory conditions.

Our standard test species among fish are cod (*Gadus morrhua*) and flounder (*Pleuronectes flesus*), one free-swimming and the other bottom-dwelling.

Among decapod crustaceans, prawn (*Leander adspersus*) and shore crab (*Carcinus maenas*), *i.e.* one more mobile and one less mobile species. Sometimes hermit crabs (*Eupagurus bernhardus*) and spider crabs (*Hyas araneus*) have also been used.

The following bivalves have been chosen: the scallop (*Pecten opercularis*), a swimming species; the cockle (*Cardium edule*), which is a superficially burrowing type; the clam (*Mya arenaria*), a deep-burrowing and relatively stationary species; and the common mussel (*Mytilus edulis*), which represents a sessile form.

Before testing, the animals are acclimatised to laboratory conditions for at least 48 hours to retain normal behaviour and physiology. If fish and crustaceans are kept for longer periods they are fed up to 48 hours prior to the test. Even if the animals are fed, they may lose their condition in the store tanks; therefore newly caught animals are generally preferred for the testing. Exceptions are species with vertical seasonal migration, such as prawns and crabs, which are caught in shallow water at the end of the summer and kept in store tanks during the winter. In long-term testing even bivalves are fed by addition of plankton twice a week.

Considering the variation of susceptibility during the life cycle, it is preferable to select test animals of the same size in order to obtain comparable results.

For the study of the effects on early phases of the life cycle, fertilisation included, eggs and larvae of cod, common mussels and lobsters have been used. The effects on the sensitive phases of the moulting cycle in decapods are studied on prawns.

BIOLOGICAL FUNCTIONS STUDIED

To determine mortality, different *criteria for death*, appropriate to each species, are used.

For fish and decapod crustaceans, the cessation of respiratory movements and body movements is used. For bivalves the cessation of function of the adductor muscles and the complete inactivation of the mantle-edge muscles are used. The planktonic eggs of cod and larvae of cod, lobster and mussels are considered dead when lying on the bottom of the tank.

To determine sublethal and chronic effects of the different materials, the following biological functions of adult animals are studied: locomotory behaviour of mobile species, respiratory movements of fish, response to food by fish and decapods, shell-closure ability of bivalves, burrowing and siphon retraction of cockles and clams, and byssal activity of common mussels. The effects on hatching rates and on the development of eggs and larvae are observed.

DATA PRESENTATION

The results obtained on *acute toxicity* and *lethal effects* for each species are presented in diagrams of toxicity or survival curves which express the time/concentration relationship for the materials tested. From these curves, the 96-hour LC_{50} values, the standard form used for the comparative ranking of toxic substances, are interpolated and tabulated.

Examples of survival curves are shown in Figs. 5.2 and 5.3, illustrating the toxicity of some dispersants to cod and common mussel, one sensitive, the other more resistant.[6] The distribution of the curves in the diagram shows the different toxicities of the tested material, and the development of oil dispersants from the older, highly toxic products to lower-toxic products can be followed. If Figs. 5.2 and 5.3 are compared, it is evident that the two species react very differently to the substances. The toxicity to cod generally decreases rapidly with decreasing concentrations, while to common mussels it decreases much more slowly, and most of the mortality occurs after exposure. This is a characteristic of bivalves which stresses the importance of including a period of clean sea water in the tests.

The 96-hour LC_{50} values of the products are tabulated and ranked for each species, as is shown in Tables I and II,[6] where the LC_{50} values obtained after the period in clean sea water are also included. Owing to the different ability of recovery of different species and the often delayed mortality, particularly of bivalves, these values are often lower than the 96-hour LC_{50}

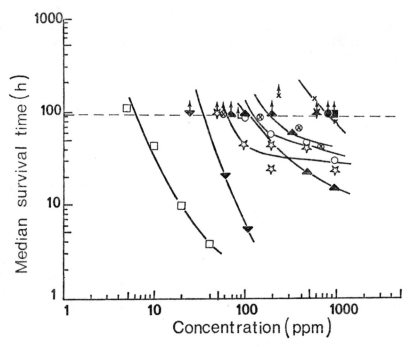

FIG. 5.2. Toxicity to cod of some oil dispersants and a surfactant at 96-hour exposure and then in clean sea water. Symbols with an arrow indicate that median mortality is not obtained. Berol TL-188 ☆, Berol TL-198 ✕, BP 1100 ▲, BP 1100X ★, Corexit 7664 ◯, Corexit 8666 ●, Fina-Sol OSR-2 ⊗, Fina-Sol SC ▼, Polyclens TS7 ■, NP10 EO □.

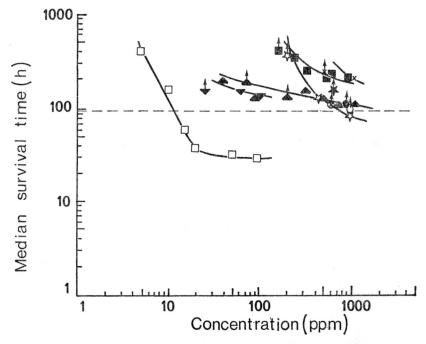

Fig. 5.3. Toxicity to common mussels of some oil dispersants and a surfactant at 96-hour exposure and then in clean sea water. Symbols with an arrow indicate that median mortality is not obtained. Berol TL-188 ☆, Berol TL-198 ✕, BP 1100 ▲, BP 1100X ★, Corexit 7664 ○, Corexit 8666 ●, Fina-Sol OSR-2 ⊗, Fina-Sol SC ▼, Polyclens TS7 ▣, NP10 EO □.

values. The tables show the different rankings of the various dispersants as to toxicity for the two species and also that the oil emulsions generally are more toxic than both the new-type dispersants and the unemulsified crude oil.

The variation obtained in survival times and in LC_{50} values for each species, probably due to temperature, season and individual condition, has been statistically treated and the 95% confidence limits calculated by Litchfield's method[2] respective to that of Litchfield and Wilcoxon.[3]

Mortality in controls has been rare. When this occasionally occurs it has been taken into consideration at the calculation of the percentage mortality of the test series as described by Tattersfield and Morris.[7]

The effects of surface-active agents, including oil dispersants, on the biological functions studied appear according to certain patterns, in the following sequence:

1. Increased activity.
2. Decreasing activity.
3. Immobilisation and death.

This sequence is accelerated with increasing concentration.

The lowest concentrations of the toxic materials which affect the different phases of this activity sequence for 50% of the animals have been calculated.

TABLE I

Median Lethal Concentrations (LC$_{50}$) for Cod in the Tested Materials at 96-hour Exposure and then after 48 hours in Clean Sea Water

Tested materials	LC$_{50}$ (ppm)	
	96 hrs	96 + 48 hrs
Surfactant:		
Nonylphenol ethoxylate[a] (NP10 EO)	6	5
Dispersants:		
Fina-Sol SC	30	30
Berol TL-188	60	60
BP 1100	120	80
Corexit 7664	130	–
Fina-Sol OSR-2	180	130
BP 1100X	>688	>688
Berol TL-198	850	600
Corexit 8666	>940	>940
Polyclens TS7	>984	>984
Oil emulsions:		
Corexit 8666 + light marine diesel 1:9	60	60
Corexit 7664 + light marine diesel 1:9	70	70
Corexit 7664 + Oman crude oil 1:1	120	120
Corexit 7664 + fuel oil 4 1:1	140	140
Oman crude oil	>1000	>1000

[a] Nonylphenol ethoxylate is a non-ionic surfactant commonly used in dispersants and detergents.

In Table III examples of such values are given for the sublethal phases of the sequence, such as swimming activity of cod, and shell-closure ability and byssus thread formation of common mussels exposed to dispersants and to oil emulsions.[6] These values are often considerably lower than the LC$_{50}$ values obtained. The rankings of the products as to their sublethal effects are also often different for the species and for the different functions studied.

RELATIONSHIP OF LABORATORY RESULTS TO THE FIELD SITUATION

The principal purpose of bioassay work to determine the relative toxicities of different materials does not mean that the results cannot also be used for the prediction of *general* ecological consequences of pollution in the marine environments, if the bioassay is given a biological design and if its limitations are kept in mind when interpreting the results to field situations.

A biological design means the use of several species representing different modes of life and divergent habitats, and belonging to various taxonomic groups. As essential as the determination of lethal toxicity is the study of

TABLE II

Median Lethal Concentrations (LC_{50}) for Cockle and Common Mussel in the Tested Materials at 96-hour Exposure and then after 48 hours in Clean Sea Water

	LC_{50} (ppm)			
	Cockle		Common mussel	
Tested materials	96 hrs	96 + 48 hrs	96 hrs	96 + 48 hrs
Surfactant:				
Nonylphenol ethoxy-late[a]	6	5	12	10
Dispersants:				
Fina-Sol SC	30	20	>110	90
BP 1100	<40	<40	>1000	250
Polyclens TS7	~178	~178	>984	>984
Berol TL-188	200	130	800	400
Berol TL-198	450	270	>1050	>1050
BP 1100X	>688	>688	>688	>688
Fina-Sol OSR-2	>700	>700	>700	>700
Corexit 8666	>940	>940	>940	>940
Corexit 7664	1000		>1000	
Oil emulsions:				
Corexit 7664 + fuel oil 4 1:1	<25	<25	>1000	>1000
Corexit 8666 + light marine diesel 1:9	50	40	>1000	>1000
Corexit 7664 + Oman crude oil 1:1	60	<50	1000	800
Corexit 7664 + light marine diesel 1:9	140	50	>1000	>1000
Oman crude oil	>1000	>1000	>1000	>1000

[a] Nonylphenol ethoxylate is a non-ionic surfactant commonly used in dispersants and detergents.

sublethal and chronic effects on biological functions important for the survival of the animals in their natural environments. Such functions can be loco-motory behaviour, defence reactions or food responses of adult animals, and growth, development and hatching rates of eggs and larvae.

The results of initial lethality and sublethal effects obtained at *short-term* testing at high concentrations may well correspond to the field situation occurring at an accidental pollution, e.g. to the acute effects of oil spills of considerable quantitites but often of short duration by quick counter-measures and rapid dilution.

From the knowledge of the lowest concentrations or threshold concentra-tions that affect the behaviour of different species in an ecosystem, important conclusions on the ecological consequences of pollution can be made. The

TABLE III

Lowest Median Concentrations of the Tested Materials Affecting Locomotory Behaviour of Cod and Shell-closure and Byssal Activity of Common Mussel at 96-hour Exposure

Tested materials	Concentrations (ppm)			
	Cod		Common mussel	
	Increased activity	Decreased activity	Decreasing shell-closure ability	Decreasing byssal activity
Surfactant:				
Nonylphenol ethoxylate[a]	2	4	10	~5
Dispersants:				
Fina-Sol SC	–	26	110	~12
BP 1100	40	~100	500	<40
Berol TL-188	50	~100	200	100
Corexit 7664	50	100	>1000	500
Fina-Sol OSR-2	–	143	>688	110
BP 1100X	288	–	700	143
Berol TL-198	–	525	>1050	100
Polyclens TS7	350	608	>984	~178
Corexit 8666	>940	>940	>940	65
Oil emulsions:				
Corexit 7664 + light marine diesel 1:9	25	25	>1000	<25
Corexit 8666 + light marine diesel 1:9	–	25	>1000	<25
Corexit 7664 + Oman crude oil 1:1	50	50	>1000	<50
Corexit 7664 + fuel oil 4 1:1	100	250	>1000	<25
Oman crude oil	~350	~350	>1000	~350

[a] Nonylphenol ethoxylate is a non-ionic surfactant commonly used in dispersants and detergents.

biological significance of increased activity is avoidance, *i.e.* flight reactions in mobile species or other protective mechanisms in sedentary species such as shell closure of bivalves. Protective mechanisms seem to be especially well developed in *littoral* species, naturally adapted to large variations of their environment. These reactions increase the chances of survival of individuals at *shorter* exposure to pollution and may explain the often slight effects of accidental pollution (oil spill) of limited size.

At a *longer* exposure, the impairment of biological functions that follows the initial period of avoidance means a disadvantage to the individual as all reactions become weaker: locomotion, response to food defence, respiration, byssal activity, etc. As different species are differently tolerant, *i.e.* have different thresholds of reaction to environmental stress, the most sensitive species are affected first if the area becomes polluted or is exposed to extreme environmental conditions. In the natural habitat, where the competition for food and space is important and where the animals are included in a prey/predatory system, the most resistant species are favoured while the most sensitive species find their chances of survival diminished in such situations. This exemplifies how a pollutant, present in a concentration that is sublethal for an individual isolated in an aquarium, becomes lethal in an ecosystem, and it shows the limitation of toxicity studies in the laboratory compared with field studies and must be kept in mind when interpreting such results to field situations.

The results obtained at prolonged exposure may explain why constant pollution, even in low concentrations, is more serious for the marine fauna and flora than a limited acute pollution. It is well known that chronic pollution disturbs the biological balance and gives rise to an ecological succession of animal and floral communities, which starts with decreasing diversity of species and ends with the replacement of metazoic communities by microbes.[4]

This points to the importance of the determination—in the field—of *ecological threshold concentrations* of the toxic materials introduced into the sea. Knowledge of such values is necessary to determine effective limits for permissible concentrations of pollutants in controlled marine environments.

REFERENCES

1. Granmo, Å., and Kollberg, S. O., A new simple water flow system for accurate continuous-flow tests, *Water Research*, **6**, 1597–9 (1972).
2. Litchfield, J. T., Jr., A method for rapid graphic solution of time–percent effect curves, *J. Pharmac. Exp. Ther.*, **97**, 399–408 (1949).
3. Litchfield, J. T., Jr., and Wilcoxon, S., A simplified method of evaluating dose-effect experiments, *J. Pharmac. Exp. Ther.*, **96**, 99–113 (1949).
4. Reish, D. J., 'A Critical Review of the Use of Marine Invertebrates as Indicators of Varying Degrees of Marine Pollution', FAO Technical Conf. on Marine Pollution and its Effects on Living Resources and Fishing, Rome, 1970, FIR: MP/70/R-9, 13 pp.
5. Swedmark, M., Braaten, B., Emanuelsson, E., and Granmo, Å., Biological effects of surface active agents on marine animals, *Mar. Biol.*, **9**, 183–201 (1971).
6. Swedmark, M., Granmo, Å., and Kollberg, S. O., Effects of oil dispersants and oil emulsions on marine animals, *Water Research*, **7**, 1649–72 (1973).
7. Tattersfield, F., and Morris, H. M., An apparatus for testing the toxic values of contact insecticides under controlled conditions, *Bull. Entomol. Res.*, **14**, 223–33 (1924).

6

Effects on Community Metabolism of Oil and Chemically Dispersed Oil on Baltic Bladder Wrack, *Fucus vesiculosus*

BJÖRN GANNING

and

ULF BILLING

(*Department of Zoology and the Askö Laboratory, University of Stockholm, Sweden*)

INTRODUCTION

Oil is known to have deleterious effects on plants and it has for a long time been employed in the control of weeds.[10] The very significant physiological effects of oil in water on plants have recently been reviewed by Baker.[1] Most seaweeds seem to suffer damage due to oil spills and succeeding clean-up, although growth stimulation has been observed in certain cases.[2,3,5] The damage may be explained by overweight of the weeds and a 'break-off' by waves, but decreases in photosynthetic capacity of attached algae have also been demonstrated.[6,12] Few quantifications of the effects on algae of oil and oil emulsions have, however, been carried out. Copeland and Dorris[7,8] used the single-diel oxygen curve method to study the effects of oil refinery effluents in holding ponds. They demonstrated an improvement on community metabolism in the effluents due to holding time and distance from the outlet. In the present investigation the method of studying algal metabolism by analyses of dissolved oxygen and its diel changes[13] has been used, correlated to concentrations of oils and dispersed oil.

MATERIALS AND METHODS

The present study was carried out close to the Askö Laboratory in the northern Baltic proper, 70 km south of Stockholm. The salinity is about 6‰ S in the area. The recordings were made on 22 June, 28 June, 16 July and 18 July 1972 when the sky was clear and the water temperature was exceptionally high, about 20°C. The annual mean water temperature in the area is about + 5°C.

Fucus vesiculosus L., one of the very few large brown weeds of the northern

53

Baltic, was collected from one locality from 1–2 m depth with a minimum of epiphytic algae (*Pilayella littoralis* (L.) Kjelm. and *Elachista fusicola* (Vell.) Aresch.). All macroscopic animals were completely removed. Algal fronds of about 25–30 g (dry weight) were placed in white 60-litre tubs, filled with 50 litres of sea water. The tubs were mounted in two rows on a raft, nine in each row, with the water bodies almost submerged. Thus, the temperature of the tubs followed the ambient water temperature $\pm$ 1°C. No changes in salinity were observed during the tests. The raft was kept in place with six long sticks pushed into the bottom of a shallow bay, behind a breakwater. The raft was floating, following changes in water level. No waves could change the chemical composition of the tub water.

Oxygen analyses were carried out every third hour with a battery-operated polarographic oxygen meter (YSI, 54RC). The accuracy of the equipment is 1% of full-scale indication. The Teflon membrane of the probe was protected against floating oil by penetrating the surface through an oil-free tube fixed in one corner of each tub. Calibration was made before and after each series of recordings to control the membrane.

The oil and oil emulsions were prepared and thoroughly mixed in the laboratory before adding to the tubs. Only wind agitation was used and irregular patches of non-emulsified oil floated on the water surface. Concentrations of 100, 560, 1000 and 5600 ppm of oil and oil + dispersant (total concentration) were used. Fuel oil 4 (supplied by BP, Sweden) and the emulsifier Corexit 7664 (Esso Chemicals) were studied (Figs. 6.1 and 6.2). A few recordings were made including diesel oil, kerosene and BP 1100X. Oxygen recordings were carried out 24 hours before and after the addition of chemicals. On one occasion (Table II) recordings were carried out 2 days after the removal of floating oil and dispersants (18 July). The recordings on 22 June were carried out after 14 days incubation in each mixture, 7 days in running sea water and 1 day of adaptation in the tubs (Figs. 6.3, 6.4 and 6.5). From these recordings the community metabolism could not be calculated.

The algae were dried at 105°C for 7 days and all calculations are reproduced per g algae dry weight.

For calculations of primary production and community respiration no corrections for diffusion were made except in the controls, although an oil cover of the water does not prevent diffusion totally.[11,14]

RESULTS

Results from test series with fuel oil 4 and fuel oil 4 + Corexit 7664 are reproduced in Figs. 6.1 and 6.2 and in Tables I, II and III. From the figures and tables it can be seen that there is an increase in community respiration, R, while the gross primary production, P, is almost independent of changes in oil and emulsion concentration. The net primary production, $P - R$, shows a smooth decrease in correlation to increasing chemical concentration (Table I). Respiration exceeds primary production between 560 and 1000 ppm of unmodified fuel oil, while even 100 ppm of fuel oil + Corexit gives higher respiration than primary production values. Fig. 6.2 also shows a confusing levelling-out of the oxygen rate-of-change line when Corexit is present. In

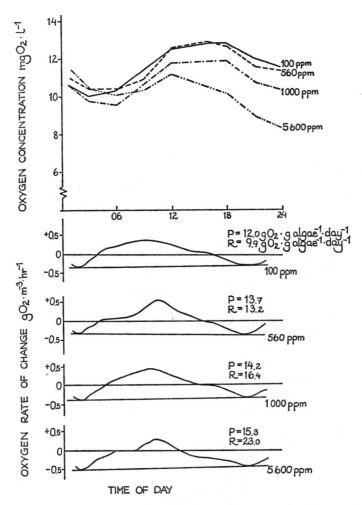

FIG. 6.1. Diel oxygen curves and productivity calculations for *Fucus vesiculosus* contaminated with fuel oil 4.

Table II recordings 2 days after removal of surface-floating chemicals are reproduced. In the tubs which had contained fuel oil a marked recovery may be noticed while all the tubs with oil + dispersant still show higher respiration than production values. The skimming-off of emulsified oil was, however, very unsuccessful. Most of the chemicals were evenly distributed in the total water volume in those tubs. The increase in net production in 5600 ppm fuel oil + Corexit may be explained simply as a matter of diffusion. Here the original oxygen concentration was 2·2 and the final concentration 3·6 mg O₂/litre, an increase which gives a positive net primary production with the method used.

A few recordings of long-time rehabilitation of *F. vesiculosus* were carried out in June. After 14 days in each mixture the weeds were kept in running

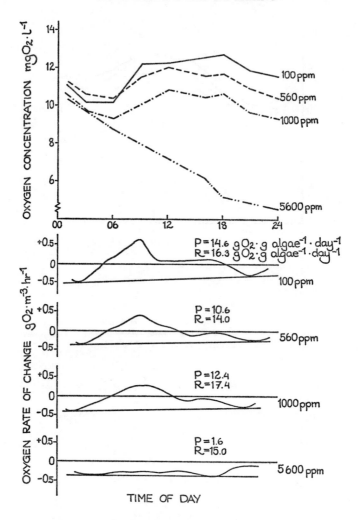

FIG. 6.2. Diel oxygen curves and productivity calculations for *Fucus vesiculosus* contaminated with fuel oil 4 + Corexit 7664.

sea water. After that time they were put in clean sea water in the tubs with an initial oxygen concentration of about 8·5 mg O_2/litre. After another 24 hours the oxygen pulse was followed for 27 hours. The *Fucus* that had been in contact with fuel oil, 1000 ppm, fuel oil + Corexit 7664, 1000 and 5600 ppm, and fuel oil + BP 1100X, 1000 ppm, showed normal diel changes in oxygen concentration. Oil + dispersant, however, gave changes on a lower level than exclusively oil. Only fuel oil + BP 1100X, 5600 ppm, gave depressed oxygen production (Fig. 6.3). Using diesel oil or kerosene + emulsifiers, only the oil and oil + Corexit 7664, 1000 ppm, gave recovery of the weeds (Figs. 6.4 and 6.5). Also in these series the emulsifier lowered the oxygen level. A subjective observation during these recordings was that there was an

TABLE I

The Immediate Effects of Oil and Oil Emulsion on Gross and Net Primary Production and Respiration of *Fucus vesiculosus* at 4 Concentrations

| Concentration | Fuel oil 4 | | |
	P	R	P − R
100	16·7	15·8	0·9
100	12·0	9·9	2·1
560	19·8	19·6	0·2
560	13·7	13·2	0·5
1000	14·2	16·4	− 2·2
5600	15·3	23·0	− 7·7

| Concentration | Fuel oil 4 + Corexit 7664 | | |
	P	R	P − R
100	14·6	16·3	− 1·7
100	13·1	16·2	− 3·1
100	14·2	14·5	− 0·3
560	12·9	18·0	− 5·1
560	10·6	14·0	− 3·4
560	8·2	13·1	− 4·9
1000	12·4	17·4	− 4·0
5600	1·6	15·0	− 13·4

increased soaking of brown algal pigments with increasing concentration of emulsifiers, especially noted in combination with diesel oil.

Table III shows that there is a fairly good reproducibility of the method used when considering net primary production while the level of *P* and *R* differs a lot in individual series.

The control series (Table IV) shows that the method used gives values well in accordance with other recordings from the Baltic and the Swedish west coast (unpublished results). However, the calculated values in Table IV may not be directly compared with those of Tables I to III because of differences in using a diffusion constant only in the controls.

DISCUSSION

The littoral organisms are very much exposed to accidental or deliberate oil spills. Much has been written about biological and ecological effects of oil spills on littoral communities,[4,9] but most of the effects have been studied in tidal areas. The Baltic Sea is cold and lacks a tidal movement but contains large and shallow archipelagos in which drifting oil is easily trapped. Owing to the extremely long shoreline, vast areas are rapidly contaminated by oil

TABLE II

Gross and Net Primary Production and Respiration of _Fucus vesiculosus_ 2 Days after Skimming off the Chemicals from the Water Surfaces

| Concentration | Fuel oil 4 | | |
	P	R	P − R
100	10·7	11·4	−0·7
100	8·9	7·6	1·3
560	11·1	3·3	7·8
560	8·5	5·7	2·8
1000	9·1	6·5	2·6
5600	14·7	9·2	5·5

| Concentration | Fuel oil 4 + Corexit 7664 | | |
	P	R	P − R
100	6·1	6·7	−0·6
100	9·1	9·7	−0·6
100	8·7	9·2	−0·5
560	8·2	15·9	−7·7
560	4·4	7·2	−2·8
560	5·5	10·9	−5·4
1000	7·3	13·9	−6·6
5600	5·2	3·8	1·4

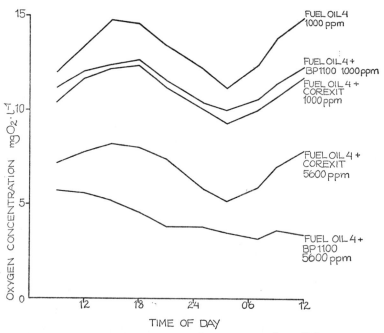

FIG. 6.3. Rehabilitation of _Fucus vesiculosus_ measured as diel oxygen curves after 14 days incubation in fuel oil 4, Corexit 7664 and BP 1100X, 7 days in running water and 24 hours in stagnant water. Initial oxygen concentration 8·5 mg/litre.

TABLE III

Gross and Net Primary Production and Respiration of *Fucus vesiculosus* in 2 concentrations of Oil and Oil Emulsion

| | Fuel oil 4 | | |
Concentration	P	R	P − R
1000	14·3	16·9	−2·6
1000	17·2	19·9	−2·7
1000	9·4	12·1	−2·7
1000	11·9	16·0	−4·1
5600	10·3	17·2	−6·9
5600	11·8	15·2	−3·4
5600	7·3	12·5	−5·2

| | Fuel oil 4 + Corexit 7664 | | |
Concentration	P	R	P − R
1000	1·7	23·6	−21·9
1000	10·1	30·0	−19·9
1000	4·9	21·7	−16·8
1000	0·9	13·9	−13·0
5600	5·5	30·9	−25·4
5600	3·4	23·5	−20·1
5600	2·0	21·0	−19·0

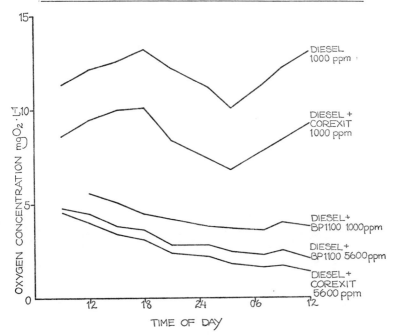

FIG. 6.4. Rehabilitation of *Fucus vesiculosus* measured as diel oxygen curves after 14 days incubation in diesel oil, Corexit 7664 and BP 1100X, 7 days in running water and 24 hours in stagnant water. Initial oxygen concentration 8·5 mg/litre.

TABLE IV

Community Metabolism in Tubs Containing *Fucus vesiculosus* and Clean Sea Water

| Date | Controls | | |
	P	R	P − R
28 June	19·2	4·2	15·0
	17·7	4·5	13·2
	18·0	3·8	14·2
	mean value		14·1
16 July	23·0	13·8	9·2
	14·3	3·9	10·4
	11·8	5·8	6·0
	mean value		8·5
18 July	24·4	13·3	11·1
	11·1	0·8	10·3
	10·3	3·4	6·9
	mean value		9·4

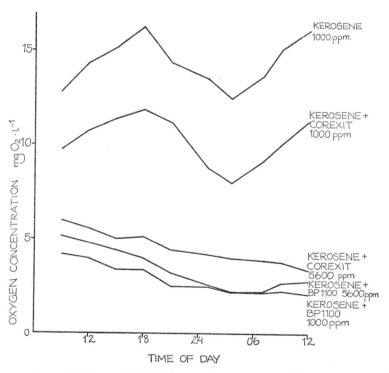

Fig. 6.5. Rehabilitation of *Fucus vesiculosus* measured as diel oxygen curves after 14 days incubation in kerosene, Corexit 7664 and BP 1100X, 7 days in running water and 24 hours in stagnant water. Initial oxygen concentration 8·5 mg/litre.

and the negative effects may be expected to be worse in the Baltic than in a tidal region with warmer water.

The dominating seaweed in the Baltic is *Fucus vesiculosus*, and within the *Fucus* belt most of the Baltic animal species live. This weed reacts with oil and emulsifiers by increasing its respiration. If the oil, even if it is diesel oil or kerosene, is removed from the water, the weed survives even 14 days of contamination at 1000 ppm while emulsified oil turns out to be much more harmful. No rehabilitation is found after 14 days when Corexit 7664 at 5600 ppm or BP 1100X at 1000 or 5600 ppm are used. Immediate decreases in net primary production are found already at 100 or 560 ppm if the oil is emulsified, while this effect is not found until 1000 ppm of fuel oil are reached. Now even the fairly high concentrations of 1000 and 5600 ppm are easily reached in sheltered areas like the Baltic archipelagos. From these results the conclusion must be drawn that shallow littoral zones in non-tidal areas should be cleaned mechanically and not by emulsifiers unless a strong current immediately carries away the oil-contaminated water.

REFERENCES

1. Baker, J. M., 'The Effects of Oils on Plant Physiology', in *The Ecological Effects of Oil Pollution on Littoral Communities* (ed. E. B. Cowell), Institute of Petroleum, London, 1971, pp. 88–98.
2. Baker, J. M., 'Growth Stimulation Following Oil Pollution', ibid., pp. 72–7.
3. Bellamy, D. J., John, D. M., and Whittick, A., The 'kelp forest ecosystem' as a 'phytometer' in the study of pollution of the inshore environment, *Underwater Ass. Rep.*, 1968, 79–82.
4. Carthy, J. D., and Arthur, D. R. (eds.), The biological effects of oil pollution on littoral communities, *Fld Studies*, 2 (suppl.), 198 pp. (1968).
5. Clendenning, K. A., The effects of waste discharge on kelp, *Fuel Oil. Univ. Calif. Inst. Mar. Resources*, 59 (4), 4–12 (1959).
6. Clendenning, K. A., and North, W. J., Effects of wastes on the giant kelp, *Macrocystis pyritera*, *Proc. Intern. Conf. Waste Dispos. Mar. Environ.*, 1, 82–91 (1960).
7. Copeland, B. J., and Dorris, T. C., Photosynthetic productivity in oil refinery effluent holding ponds, *J. Water Pollut. Contr. Fed.*, 34, 1104–11 (1962).
8. Copeland, B. J., and Dorris, T. C., Community metabolism in ecosystems receiving oil refinery effluents, *Limnol. Oceanogr.*, 9, 431–47 (1964).
9. Cowell, E. B. (ed.), *The Ecological Effects of Oil Pollution on Littoral Communities*, Institute of Petroleum, London, 1971, 250 pp.
10. Currier, H. B., and Peoples, S. A., Phytotoxicity of hydrocarbons, *Hilgardia*, 23, 155–73 (1954).
11. Mortimer, C. H., The exchange of dissolved substances between mud and water in lakes, *J. Ecol.*, 20, 280–329 (1941).
12. North, W. J., Neushul, M., Jr., and Clendenning, K. A., Successive biological changes observed in a marine cove exposed to a large spillage of mineral oil, *Symp. Comm. Int. Explor. Sci. Mer. Médit.*, Monaco, 1964, pp. 335–54.
13. Odum, H. T., Primary production in flowing waters, *Limnol. Oceanogr.*, 1, 102–17 (1956).
14. Ottway, S. M., 'Oil Films and Gas Exchange', in 'Some Effects of Oil Pollution on the Life of Rock Shores', M.Sc. thesis, University of Wales, 1972.

7

Toxicity Testing at the Station Marine d'Endoume

GERARD L. BELLAN

(Station Marine d'Endoume, Marseille, France)

Since 1968 an important effort has been devoted at the Station Marine d'Endoume to improve methods employed in the study of the action of several pollutants, especially emulsifiers, detergents and allied substances, on some marine invertebrates.

The first step of the research has consisted of a series of short-term bio-assays, of a few days duration; later a second step has been developed, with long-term bioassays over a period of a few weeks.

As test animals, we have tried to choose species representative of various groups of invertebrates (polychaetes, molluscs and crustaceans), and also to represent different trophic levels and feeding types (herbivorous, omnivorous, suspension feeders, detritus feeders, carnivores, etc.), with well-defined ecological requirements, living in biotopes exposed to pollution the nature and the intensity of which is different. We included species living in very clean water, and some not tolerating any pollution.

I. SHORT-TERM TESTS

We use only specimens sampled from the natural environment, in order to obtain a better estimation of the influence of pollutants on natural popula-tions. We have collected these specimens as far as possible from within homogeneous populations and always from the same locality for each species.

After collection, animals are acclimatised over several (5–7) days in a constant-temperature aquarium with running water. Individuals which are not in good condition are eliminated. A few hours before experimentation, animals are stocked in the experimental room at 17°C to reduce possible stress due to a temperature change.

The experimental solutions are prepared with offshore sea water without detergents, kept some weeks in a dark room in the laboratory. The detergents used for the toxicity tests are dissolved in this sea water.

(a) Lethal Time 100% Tests (LT_{100})[1,6,8]

During the first stage of our research these tests were conducted with the following species: polychaetes (*Scolelepis fuliginosa*, *Capitella capitata*), molluscs (*Mytilus galloprovincialis*, collected from unpolluted, little-polluted or heavily polluted waters) and crustaceans (*Sphaeroma serratum*, *Idotea balthica basteri*, *Gammarus olivii*).

Generally speaking, we use 12 different concentrations: 800, 400, 200, 100, 50, 25, 10, 5, 2·5, 1, 0·5 and 0·1 mg/litre, plus a control (sea water without detergent). (It is also possible to have another range of concentrations: 1000, 100, 10, 1, 0·1 and 0·01 mg/litre, plus a control.) As all these animals are able to tolerate a few days starvation, they are not fed.

Polychaetous annelids are placed in a 0·370-litre glass jar with 0·175 litre of solution for 96 hours, molluscs in 4-litre white polyethylene bowls with 2 litres of solution for 192 hours, crustaceans are put in a glass jar with 0·5 litre of solution for 96 hours.

Dissolved oxygen measurements have convinced us that this factor was not very important (in the special case of our experiments), so we have not aerated our experimental media; artificial oxygenation provokes a decrease of the toxicity (up to 50%).

Mortality criteria have been determined for each species.

In general, and according to some recommendations of the Zürich Convention, we have used ten animals for each test concentration. These ten specimens are placed in a container (it seems better to separate each animal in a particular jar, but this requires a very large constant-temperature room which was not at our disposal).

Observations are made at least twice a day, and the number of dead animals recorded with the time of observations. These dead animals are picked out of the containers. Tables and mortality curves are then drawn up.

(b) Lethal Concentration 50% Tests (LC$_{50}$)[10]

In a work the scope of which is a complete study of the determination of lethal concentrations (LC$_{50}$ or other lethal concentrations) of toxic products in experiments using marine animals, a part of the experimental technique used for the polychaetes *Scolelepis fuliginosa* and *Capitella capitata* has been modified. The LC$_{50}$ was determined by three different experimental methods: to establish the concentration where 50% of the animals are killed, (i) by the mathematical method of Bliss,[5] (ii) using the probit transformation and (iii) by the graphic method.

Twenty animals are used for every concentration, plus a control. The animals are placed in polystyrene Petri dishes with 50 ml of solution, with only one animal per dish (to avoid self-pollution in containers). The animals are observed after 24, 48, 72 and 96 hours. In a first experiment, ten concentrations are used: 800, 400, 200, 100, 50, 25, 10, 5, 2·5 and 1 mg/litre, plus control. This experiment makes it possible to establish the concentration limits between which the LC$_{50}$ is located. In a second experiment a scale of concentrations has been established between the limits obtained in the first experiment. For example, if we find an LC$_{50}$ 48 hours between 10 and 5 mg/litre in the first experiment, in the second experiment we use 10, 9, 8, 7, 6 and 5 mg/litre plus a control. From this second experiment an evaluation arises that is more accurate of the mortality than in the first case, allowing the experimental determination of the required LC$_{50}$.

The computation of LC$_{50}$ according to Bliss's mathematical method[5] is then possible. The transformation of a sigmoid curve to a straight line also permits determination of the LC_{50} from a graph.

The results from the Bliss method are more accurate than the experimental results, which in any case give two limit values between which the real lethal concentration is placed. The test by the Bliss method also tests the homogeneity of the sample of organisms used in the experiment.

II. LONG-TERM TESTS[2,4]

Studies *in vitro* of the effects of detergents on marine organisms have been largely concerned with short-term lethal effects. Sublethal long-term tests obtain quantitative data on the effects of detergents on various stages of the life cycle from the appearance of eggs, through reproduction to the early benthic stage of the next generation. These tests also give some information about morphological and physiological effects of pollutants.

The concentrations employed are 1000, 100, 10, 1, 0·1 and 0·01 mg/litre, plus a control, with the polychaetes *Capitella capitata* and *Scolelepis fuliginosa* and the isopod *Idotea balthica basteri*. To date, the species most studied is *Capitella capitata*.

For this species, we use a set of five or ten 0·5-litre glass jars with 0·45 litre of test solutions. Five males and five females, 30–40 days old, are placed in each container (it would be better to isolate one couple in each glass jar). The animals in each experimental container are fed with about 5 mg dried *Ulva* every five days. The jars are placed in a constant-temperature room (17°C). Observations are made twice daily and the time completion of each stage is recorded. The different stages in the life cycle of *Capitella capitata* or the various periods are as follows:

1. Appearance of ovarian tissue; the time necessary for the eggs to mature is noted.
2. Female spawning; the time of egg-laying within the maternal tube is recorded (with the number of eggs).
3. Incubation period; the time necessary for development up to the trochophore stage is recorded.
4. Trochophore stage with two phases; trochophores into the tube and free trochophores outside.
5. Metatrochophore stage.
6. Young benthic stage.

The time required to reach each stage is recorded.

After a female has laid her eggs, she is placed in a small glass container containing 3 ml of test solution and the number of eggs is counted under the binocular microscope. Afterwards this female, with developing eggs, is returned to the same experimental container. At the end of the incubation period the trochophore larvae in 450 ml of sea water are counted, using 3-ml aliquots of test solution at a time. Each 3-ml solution and larvae are then placed in a clean 500-ml jar with food. When the trochophores begin to swim near the bottom, this method of counting larvae is repeated. The validity of this procedure for counting developing worms was tested by comparing the results with those obtained in a separate experiment in which each stage was counted as above, then killed and counted: the experimental variation was 5% or less. The validity of this experimental approach is that it allows the study of the effect of sublethal toxic solutions throughout the whole of the life cycle of a species and gives quantitative results at each stage.

With *Scolelepis fuliginosa* the method is slightly different: this species is fed with an artificial diet (Tetramin). Females can spawn five times during their whole life; eggs are laid in a mucoid capsule.

REMARKS

A methodology should be practical, reproducible, accurate and inexpensive.

We use stagnant mediums because experimental containers with running water require too much space. For larvae (and even small species) it seems to us necessary to have a medium without circulation.

We had no problem with dissolved oxygen, so we did not use artificial oxygenation (air bubbles, pieces of algae, etc.). With short-term tests especially it is better to avoid, as far as possible, all uncontrollable variables in the experimental conditions.

During LT_{100} tests we do not compare mortality with control, but we can accept a certain percentage of mortality if we use the Bliss method of computation for LC_{50} tests.

It is technically more correct to separate animals (or pairs), one for each experimental container, but this method increases extensively the laboratory space required and the number of manipulations involved.

REFERENCES

1. Bellan, G., Caruelle, F., Foret-Montardo, P., Kaim-Malka, R., and Leung Tack Kit, Contribution à l'étude de différents facteurs physicochimiques polluants sur les organismes marins: action des détergents sur la Polychète *Scolelepis fuliginosa* (note préliminaire), *Tethys*, **1**(2), 368–74 (1969).
2. Bellan, G., Foret, J. P., Foret-Montardo, P., and Kaim-Malka, R., 'Action *in vitro* de détergents sur quelques espèces marines', in *Marine Pollution and Sea Life*, Fishing News (Books) Ltd, London, 1972 (and FAO Technical Conference on Sea Pollution, Rome, 9–18 Dec. 1970).
3. Bellan, G., Reish, D. J., and Foret, J. P., Action toxique d'un détergent sur le cycle de développement de la Polychète *Capitella capitata* (Fab.), *C.R. Acad. Sc. Paris*, **272**, 2472–9 (1971).
4. Bellan, G., Reish, D. J., and Foret, J. P., The sublethal effects of a detergent on the reproduction, development and settlement in a polychaetous annelid, *Capitella capitata*, *Mar. Biol.*, **14**(3), 183–8 (1972).
5. Bliss, C. I., The calculation of time–mortality curve, *Ann. Appl. Biol.*, **24**, 815–52, (1937).
6. Foret-Montardo, P., Etude de l'action des produits de base, entrant dans la composition des détergents issus de la pétroléochimie, vis-à-vis de quelques espèces marines, *Tethys*, **2**(3), 567–614 (1970).
7. Kaim-Malka, R. A., Action *in vitro* des détergents non ioniques sur l'Isopode Valvifère *Idotea baltica basteri* Audouin 1827, *Tethys*, **4**(1), 51–62 (1972).
8. Kaim-Malka, R. A., Action *in vitro* des détergents non ioniques sur l'Isopode *Sphaeroma serratum* (Fabricius), *Tethys*, **4**(3), 587–96 (1972).

9. Kaim-Malka, R. A., Action *in vitro* de quelques détergents cationiques sur trois espèces de Crustaces, *Tethys*, **5** (1) 125–8 (1973).
10. Stora, G., Contribution à l'étude de la notion de Concentration léthale limite moyenne (CL_{50}) appliquée a des Invertèbres marins: 1. Remarques méthodologiques, *Tethys*, **4**(3), 597–644 (1972).

8

A Method for Testing the Toxicity of Suspended Oil Droplets on Planktonic Copepods Used at Plymouth

MOLLY F. SPOONER

and

C. J. CORKETT

(*Marine Biological Association of the United Kingdom, Plymouth, England*)

INTRODUCTION

The sensitivity of planktonic organisms to oil, whether naturally or artificially dispersed, is of obvious interest. The method described here enables an even dispersion of oil droplets to be continuously available by using vessels undergoing slow inversion on a wheel. Faecal pellet counts were used as a measure of activity, and the effects seen at the concentrations and times of exposure chosen were usually sublethal, survivors showing good recovery of feeding rate. The toxic effects of the oil may be operative in two ways, as solutes or actually ingested as droplets. Either of these may have a narcotic effect and possibly other consequences. A fuller account of this work is being prepared for publication.

APPARATUS

A wooden disc 80 cm in diameter carried a maximum of 80 specimen tubes, 35 ml, capped with coverslips and held in position with Terry clips, in three concentric rings of 16, 40 and 24 tubes. Between those of the outermost ring were 24 bottles, 200 ml, with ground-glass stoppers, held by large Terry clips and supported radially with brackets and with half clips pressing against the stoppers. The larger bottles were suitable for individual *Calanus helgolandicus* (Claus) while the tubes served at other times for smaller species of copepods. The wheel was rotated in a vertical plane at about 25 revs per hour by a $\frac{1}{50}$ h.p. motor driving through appropriate gears.

METHOD

The oil used in these experiments has been 250°C residue of Kuwait crude oil, regarded as more or less equivalent to a one-day weathered oil. Suspensions were made either by vigorously shaking with dispersants of low toxicity (5:1) or by emulsifying oil alone in sea water by pumping through a fine jet of a hand-operated emulsifier. Droplets ranged from 1–2 μ to 5–10 μ, with occasional larger droplets. The actual concentration of oil in the final dilution is checked by extraction with cyclohexane and spectrofluorimetry, concentrations of $c.$ 2 ppm and 10 ppm being normally used.

The algal flagellate used as food (*Platymonas suecica* Kylin (old name *Tetracelmis suecica* (Kylin) Butch.)) has a cell size of $c.$ 4 μ × 10 μ. The concentration was determined on a Coulter counter and adjusted to provide 120 000 cells/ml. At this concentration of algae and with oil at 10 ppm there were approximately similar amounts of oil droplets at $c.$ 4–10 μ diameter and algae, both being a suitable size for filter-feeding copepods and being collected non-selectively. The presence of the oil droplets provided more material from which to make faecal pellets, so that if the oil were entirely inert a greater number of pellets would be expected than from algal suspensions alone. A trace of this effect was sometimes seen in the lower oil concentrations, but it was not considered practicable to adjust the algal numbers to allow for the oil droplet numbers as these are so difficult to determine with any accuracy and vary so much in size according to method of preparation.

The sea water used throughout the experiments was filtered through 0·2 μ membrane and well shaken to ensure good initial oxygenation (see below for risks of reduction of oxygen content during experiments).

Copepods were collected in tow-nets from near the entrance to Plymouth Sound, overcrowding was avoided and they were kept cool during sorting. The adults of one or two predominant species were selected, sexed and fed on algal flagellates. Experiments with small copepods were normally set up one day after collection so that only fully active individuals are used. The following have so far been used singly per tube: *Temora longicornis* (O. F. Müller) males and females (1·5 mm), and *Centropages typicus* Krøyer, males and females (1·75 mm); *Acartia clausi* Giesbrecht (1·1 mm), being inconveniently small to handle singly, has been used at about 5 per tube. *Calanus helgolandicus* females (3 mm) require the larger bottles; stocks of this species remained in good health for 1–2 weeks when well fed in 1-litre volumes at 12°C so that several experiments could be made from the same stock. There was considerable variation between individuals in their feeding rates even among healthy controls and there could be variation between successive batches of animals. Several groups of comparisons of toxic suspensions were therefore made using 8 or 10 individuals per type and concentration of toxin. (Larger numbers would have been desirable, but the time required for sorting and handling becomes limiting and for *Calanus* only 24 bottles could be held on the wheel at one time.)

Copepods were gently transferred to 2 ml of sea water in each of a batch of tubes which were then filled to the brim with the required mixture of sea water and algal food for the control, or with the addition of oil suspensions to give concentrations of about 2 ppm or 10 ppm. These suspensions were kept well mixed by frequent inversions of the flask while being used to fill the tubes or bottles, any air bubbles being given time to rise. A light placed underneath meanwhile drew the copepod down while the round coverslip was dropped on. No special seal was required. Spare liquid was dried off and the tube clipped on the wheel with minimum delay. Provided the bulk suspensions were at room temperature, 12°C, air bubbles rarely developed. Such bubbles may adversely affect the behaviour of the copepod according to the experience of other workers with *Calanus* in closed, vertically rotating vessels.[1] Also oil tended to collect at any air/water interface; otherwise the oil remained very largely in even suspension, though 5–8% was found to have become adsorbed on bottle walls when checked in one experiment. Oil thus adsorbed would continue to contribute solutes and the concentration of droplets would in any case fall very slightly during the course of the experiment owing to removal by feeding.

Experiments were carried out in the dark to encourage a high feeding rate.[2] The duration of exposure to toxins has been 20 hours. The contents of the tubes were then poured into small dishes, the animals removed for recovery and the faecal pellets counted. The pellets produced by *Calanus* were reclaimed from the bottles by allowing them to stand for 30 mins, siphoning off most of the water, pipetting out the *Calanus* to a small washing dish prior to recovery, and then tipping and rinsing out the remaining water and pellets into a small dish for counting. Checks were made for stray pellets in the washing water and in that siphoned off, which was sometimes found to contain several floating pellets unusually rich in oil. Pellets produced in the controls and in the 2 ppm oil sank readily; those derived from 10 ppm suspensions sank more slowly, having near-neutral buoyancy; when examined by fluorescence microscopy these latter seemed to consist of about half oil and about half algae. The average faecal pellet production formed the numerical basis for comparing toxic effects.

After the copepods had been exposed to the various suspensions they were subjected to a recovery procedure. The preferred method was to put up to 10 animals into 500 ml sea water + algae for a week and then to test the rate of faecal pellet production by individual survivors from each type of treatment offering all animals standard oil-free algal suspension in tubes or bottles as before.

Since copepods normally show considerable and rather erratic variations in swimming activity even when healthy, only major differences in activity level have been noted. A few individuals died in the most concentrated oil suspensions, a few others were reported as slow or very poor, such individuals usually dying within 2–3 days in the 'recovery' vessel. Completely motionless animals could show some mobility after a few hours in an open dish but never recovered completely. The majority of animals were, however, only slightly less active at the end of 20 hours and their recovery was good. Those from the oil suspensions were easier to catch for transfer than those from sea water.

Oxygen conditions in the 35-ml closed tubes were tested by micro-Winkler technique, the algae being filtered off during withdrawal of the sample. Only 'controls' were tested because the presence of oil as well would have interfered with the analysis. The oxygen content was analysed after 20 hours and the difference between tubes containing algae alone and tubes containing one or five animals suggests that a fall of 2–4% of the oxygen content might be attributed to a single *Temora longicornis* or *Centropages typicus;* these figures are near the limit of accuracy of the method but are in reasonable agreement with published figures.[3] In order to obtain a larger fall in oxygen content, about 120 *Acartia clausi* were crowded in one 35-ml tube of sea water with algae. After 20 hours they were all still very active, though the oxygen had fallen to about 30% of saturation. Rates of oxygen uptake by *Calanus* females[4] at 15°C was 0·58 μl/copepod/hr and at 10°C was 0·4 μl/copepod/hr. Thus, in a bottle of 200 ml of oxygen-saturated sea water at 12°C, a fall of less than 1% might be expected in 20 hours. Algal respiration would account for a further small loss. It therefore seems that oxygen deficiency is unlikely to be affecting activity under the present experimental conditions.

Nevertheless it is admitted that the copepods, even singly, appeared not to be quite so active in the controls at the end of the 20 hours, some slight stress being imposed by the confined conditions, *e.g.* presence of obstructing glass walls. If much larger volumes per individual were to be used it would be very difficult to set up sufficient replicates and the number of simultaneous comparisons would be severely restricted. The rate of rotation is thought to be slow enough; a faster rate is detrimental.

RESULTS

Data were accumulated as sufficient copepods of a single species and sex became available. An example of results from four successive experiments from one batch of *Calanus* females, used 4–7 days after collection, is given in Table I.

In Table I each individual experiment showed a markedly slower rate of faecal pellet production in 10 ppm of oil compared with that in 2 ppm or in the control. Apart from a few mortalities there was good recovery. The much higher feeding rate seen after the week's recovery compared with the 56·5 pellets produced in the control could well be due to further acclimatisation to the food regime.

In numerous experiments with smaller copepods, similar but in general more marked effects were seen. These concentrations of oil were fortunately chosen near the critical level for detection of sublethal effects on the rate of feeding.

DISCUSSION

There is a dearth of information concerning the concentration of oil in waters under or beside slicks. One example was obtained in Tarut Bay where natural dispersal in shallow water close to a slick of week-old light Arabian crude produced droplets at 50 ppm, but wind-produced currents were causing rapid dissipation of the yellowish sea water.[5] After the *Arrow* spill 0·02 ppm or less of bunker oil particles were found, away from slicks, in a survey of Chedabucto Bay.[6] Under experimental slicks of Kuwait crude, where there was no artificial and little natural dispersal, oil was present at 0·01–0·02 ppm with occasional samples up to 0·4 ppm.[7] In samples taken one month after the *Torrey Canyon* spill where detergent had been used, values were recorded of 0·007–0·014 ppm estimated as Kuwait crude.[8] It would therefore seem

TABLE I

Effect of Kuwait 250°C Residue Oil + BP1100X on *Calanus helgolandicus* Females

(4 experiments with 8 females singly per control or type of toxin used)

Type of suspension	Total number of animals				Number of faecal pellets per active female per 20 hrs				
	Start	Active	Poor	Dead	Experiment				Mean
		After 20 hrs			1	2	3	4	
Toxic Exposure									
I. control sea water + algae	32	31	1	—	40·2	70·7	49·4	65·7	56·5
II. 2 ppm oil + 0·4 ppm dispersant + algae	32	30	2	—	59·6	74·3	61·2	53·6	62·2
III. 10 ppm oil + 2 ppm dispersant + algae	32	28	4	—	18·7	41·7	26·4	22·7	27·4
After 7 Days' Recovery									
from group I. in sea water + algae	32	31	—	1	95·6	84·1	88·3	89·6	89·4
II. in sea water + algae	31	28	2	1	102·6	106·9	112·1	93·4	103·7
III. in sea water + algae	27	26	1	—	99·5	92·6	95·9	74·8	90·7

that under conditions at sea neither the droplets nor the derived solutes are likely to persist at concentrations of 10 ppm for more than an hour or two except where being continuously produced close to a slick. It should also be remembered that zooplankton undergo diurnal vertical migration which would carry an individual out of a region of dense droplets in a few hours at the most. Thus, the slight effects seen in the present experiments of exposures for 20 hours to 2 and 10 ppm of weathered oil suggest that in open sea conditions it would be unlikely that copepods would be damaged by the amount of weathered oil which they might ingest. After the *Arrow* spill, healthy copepods with ingested oil and oily faecal pellets were collected.[9] Likewise, after the *Torrey Canyon* spill copepods appeared healthy when collected; normal patchiness makes observation of abundance difficult to interpret, but there was no reason to suspect any general damage (except that reported to floating pilchard eggs, probably caused by highly aromatic detergent).[10] Some product oils, or fresh toxic crude oils, in a fairly confined area could produce some detrimental effects on plankton; so far all experiments on the wheel have employed a less toxic 'weathered' oil. Long-term, low-level effects, which cannot be studied with this apparatus, have yet to be considered.

ACKNOWLEDGEMENTS

A vertically rotating wheel system has been used for oil pollution studies at the Institut Royal des Sciences Naturelles de Belgique by Mlle C. van der Wielen, who described it as 'less unsatisfactory than most methods for keeping oil in suspension'. The basic idea is gratefully acknowledged. The present apparatus was designed for plankton studies and constructed under the direction of Mr F. G. C. Ryder in the MBA workshop. Mrs A. Badrian gave valuable technical assistance with experiments and Dr. E. D. S. Corner made numerous helpful suggestions.

REFERENCES

1. Corner, E. D. S., Marine Biological Association of the UK, personal communication.
2. Marshall, S. M., and Orr, A. P., *Biology of a Marine Copepod*, Oliver & Boyd, Edinburgh, 1955, p. 111.
3. Marshall, S. M., and Orr, A. P., *J. Mar. Biol. Ass. U.K.*, **46**(3), 513 (1966).
4. Marshall, S. M., Nicholls, A. G., and Orr, A. P., *J. Mar. Biol. Ass. U.K.*, **20**(1), 1 (1935).
5. Spooner, M. F., *Mar. Pollut. Bull.*, **1**(11), 166 (1970).
6. Forrester, W. D., *J. Mar. Res.*, **29**(2), 151 (1971).
7. Ministry of Defence (Navy Department), *The Effects of Natural Factors on the Movement, Dispersal and Destruction of Oil at Sea*, Select Committee Report, 26 July 1968, para 61 (iii), Ministry of Defence, London, 1971.
8. Smith, J. E. (ed.), '*Torrey Canyon*' *Pollution and Marine Life*, Cambridge Univ. Press, 1968, p. 34.
9. Conover, R. J., *J. Fish. Res. Bd Canada*, **28**(9), 1327 (1971).
10. Smith, J. E. (ed.), ibid., pp. 27, 31.

9

Toxicity Testing at the Biologische Anstalt Helgoland, West Germany

WILFRIED GUNKEL

(*Biologische Anstalt Helgoland, Meeresstation, West Germany*)

INTRODUCTION

The increasing pollution of the marine environment by man-made substances which exhibit a damaging effect upon plants, animals and micro-organisms makes it necessary not only to monitor the changes occurring in ecosystems due to toxic substances, but also to predict the effects of dumped industrial wastes, effluents, sewage and spillages as well as the effects of chemicals such as dispersants used to clean up beaches and coastal waters. Administrators and politicians need information within a short time about the ecological effects of pollutants, which can be used as a help in management decisions.

The almost exclusive method used in determining toxicity is to establish the concentration where 50% (LC_{50}) of the organisms investigated are killed. The 'Interim Toxicity Procedures'[9] use the term TL_{50} and define it as 'the concentration at which just 50% of the test animals are able to survive for the specified period of exposure. This corresponds to the TL_m designation formerly used.' Likewise used in toxicity testing is the term LD_{50}, which in a more recent workshop[5] about marine environmental quality is described as 'acute toxicity-testing experiments in which a limited range of test organisms has been subjected to toxicants for a certain arbitrarily chosen time period and the dose required to kill 50%'. This method makes it possible to get results within a few days and to compare the relative toxicity of different materials. The results received by the LD_{50} method are subject to many criticisms and many discussions in scientific meetings and publications. In the above-mentioned workshop some of the shortcomings of the traditional approaches are summarised, *e.g.*

1. A limited range of test species is used.
2. Hardier laboratory organisms are used, which are likely to be less sensitive to pollutants.
3. Adults are used rather than the generally more susceptible young stages.
4. The influence of long-term sublethal concentrations on behaviour, survival, reproduction and community structure has not been determined.

The authors mention that 'the emphasis of LD_{50} data results in part from legal requirements for evidence that a given concentration will kill organisms'.

Although the criticism is justified, it is not possible to reject the LD_{50} method before other and better methods are available and generally accepted. In the above-mentioned publication a conceptual framework and specific recommendations are given about the possible improvement of toxicity testing at five different levels, stretching from the cellular level to community dynamics and structure under natural conditions.

A basic prerequisite for this is to accumulate much more information about the different influences of pollutants upon organisms and ecosystems of the sea, which is one of the goals of the work of the Biologische Anstalt Helgoland.

The Biologische Anstalt Helgoland (BAH) is a Federal research institution under the authority of the Ministry of Science and Education (since 1 January 1973 under the authority of the Ministry of Research and Technology). It was founded in 1892 and now has a staff of 133 scientists, technicians and seamen. The BAH is situated in Hamburg, Helgoland and List/Sylt and consists of the following five departments:

1. Marine Zoology.
2. Marine Botany.
3. Marine Microbiology.
4. Biological Oceanography.
5. Experimental Ecology.

The research programme of the institute covers nearly all fields of marine biology. The main emphasis is put on inter-disciplinary work to study the ecology of marine organisms with its complex mutual relationships and the importance of organisms in cycling organic matter.

The Biologische Anstalt Helgoland does not conduct routine toxicity testing of marine pollutants. This is done in Germany by, e.g., the Bundesforschungsanstalt für Fischerei and the Bundesanstalt für Gewässerkunde, but about one-third of our research capacity is dedicated to questions which are closely related to the pollution of the sea. Some of these activities are listed below:

1. Investigations about the influence of pollutants upon micro-organisms, phyto- and zooplankton and different growth stages of fishes.
2. Monitoring important environmental parameters.
3. Investigations about the occurrence and the decomposition of mineral oil by marine micro-organisms.
4. Cultivation experiments of different members of the food chain.
5. Organising international symposia, e.g. 'Biological and Hydrographical Problems of the Water Pollution of the North Sea and Adjacent Seas' (1967) and 'Cultivation of Marine Organisms and its Importance for Marine Biology'. (1969) In both symposia more than 200 scientists from all over the world participated.
6. Furnishing working space for visiting scientists who conduct investigations on different fields of marine biology and applied sciences.

ORGANISMS AND POLLUTANTS INVESTIGATED

In this paper it is not possible to review all investigations of the BAH which deal with marine pollution. Only a small selection of recent work is presented where the following organisms were used:

1. Pure cultures of the bacterium *Serratia marinorubra*.
2. Bacterial populations of freshly sampled sea water.
3. Unialgal cultures of *Prorocentrum micans* and *Ceratium furca*.
4. Fertilised eggs and larvae of the herring (*Clupea harengus*).

The effects of the following substances are described:

1. 'Red mud' ('Rotschlamm'). Waste of aluminium factories. Strong alkaline red paste, rich in iron and aluminium oxides. The possibility that millions of tons per year may be dumped in the North Sea has been discussed recently.
2. Industrial waste-water of a titanium factory which consists mainly of a mixture of diluted sulphuric acid and iron sulphate.
3. Different emulsifiers (Hüls SLM, Hüls TM and Esso Corexit).
4. Mixtures of the emulsifier Moltoklar with Iraq crude oil.

EXPERIMENTS USING UNICELLULAR ORGANISMS

Experimental Series No. 1

The influence of three different emulsifiers upon a pure culture of the marine bacterium *Serratia marinorubra* was checked. The emulsifiers used are described as 'non-toxic'. They were applied at concentrations of 100 mg/litre and 100 g/litre. The latter concentration is extremely high and is chosen only for theoretical reasons. The emulsifiers in question were:

1. Hüls SLM (mixture of carbon acid polyglycolethers and alkylolamines).
2. Hüls TM (alkylarylpolyglycolether).
3. Esso Corexit 7664.

Bacteria (stationary phase) and emulsifiers were added to sterilised aged sea water and stored for 7 days at 18°C. Aliquots were taken immediately (time 0) and after 8 hours, 1 day, 2 days and 7 days. The number of the bacteria was determined using the agar pour plate method (colony count). The results are plotted in Fig. 9.1. Bacterial numbers are plotted as logarithms against time. K stands for control determinations without emulsifier. Whilst the numbers of the control experiment did not change, all three emulsifiers at the concentration of 100 mg/litre lead to an increase of bacterial numbers, most likely serving as carbon and energy source for the growth (multiplication) of the bacteria. The addition of 100 g/litre resulted in the case of TM in a decrease of about three orders of magnitude, in the case of Corexit only a slight decrease occurred, whilst for SLM an increase of about three orders of magnitude was measured. SLM is described as a readily degradable emulsifier. Besides the fact that concentrations of 10% of an emulsifier will never occur during the application of these chemicals, it seems most likely

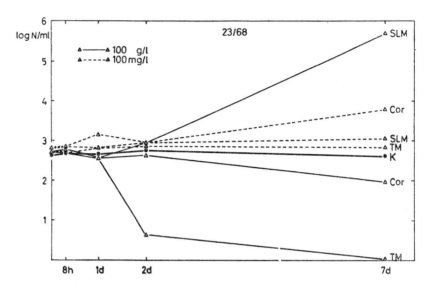

FIG. 9.1. Influence of two different concentrations of the emulsifiers Hüls SLM,
Hüls TM and Esso Corexit (Cor) upon a pure culture of *Serratia marinorubra*.
The logarithms of bacterial numbers per ml are plotted against the time of storage
at 18°C. *K* = control with no emulsifier.

that the decrease of TM is not due to the emulsifier itself but to the decrease
in pH. The manufacturer describes a 1% concentration of TM in water as
giving a pH of 4·6. For the 10% concentration of TM in the sea water used in
our experiments we measured a pH of 4·83. The corresponding pH of the
10% SLM/sea-water mixture gave a pH of 6·08.

The results show that the three emulsifiers had practically no toxic action
against the test strain of *Serratia marinorubra*.

In Fig. 9.1 only two concentrations of the three emulsifiers are plotted at
different times. Fig. 9.2 complements these findings. Here at one time of
storage (7 days) six different concentrations are plotted. *A* stands for the
original number of the test culture, *B* = emulsifier Hüls SLM, *C* = Esso
Corexit and *D* = Hüls TM. It can be seen that up to a concentration of
10 g/litre there is only a moderate influence upon *Serratia marinorubra*. At
100 g/litre the influences are already discussed in connection with Fig 9.1.

Experimental Series No. 2

To extend these experiments, the response of bacterial populations of freshly
sampled sea water against different concentrations of Hüls SLM emulsifier
was tested. The results are plotted in Fig. 9.3. Curve *A* shows the increase of
bacterial numbers of untreated sea water with the storage time at 18°C.
Owing to the 'solid surface effect'—that is, the accumulation of dissolved
organic matter present in the sea water which can now be used as a food source
by the bacteria—and dead plankton organisms and detritus, the numbers
increased. To exclude most of the particulate matter, another part of the sea
water was filtered through sterilised filter paper (average retention grade)

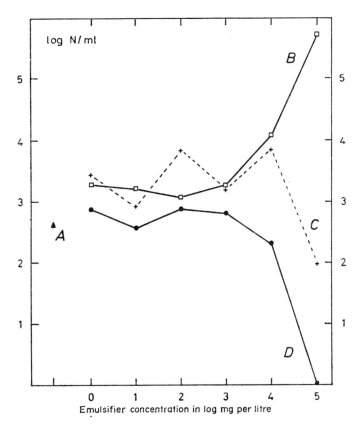

FIG. 9.2. Influence of six different concentrations of three emulsifiers upon a pure culture of *Serratia marinorubra* after 7 days of storage (for explanation, see text).

and stored at 18°C. Owing to the retention of some part of the bacteria and most of the plankton and detritus, the numbers shown in curve *B* are always lower than in curve *A*.

The curves *C*, *D* and *E* represent the data for paper-filtered sea water supplemented with: 1000 mg SLM/litre (*C*), 100 mg SLM/litre (*D*) and 10 mg/litre (*E*). In this experiment the bacteria are much more sensitive to the emulsifier than in the experiment with the pure culture. About 90% of the cells are inactivated at concentrations of 1000 mg/litre and 100 mg/litre at time 0, whilst at a concentration of 10 mg/litre there was almost no influence. During storage at 18°C the numbers of cells for all samples with emulsifiers are below the curve for no emulsifier up to 24 hours. After 48 hours, however, all three samples with emulsifiers gave higher values. The highest value was received at a concentration of 100 mg/litre, the lowest with a concentration of 1000 mg/litre. The 10 mg/litre value was between these results.

The results show:

1. The naturally occurring bacteria of the sea water were more sensitive than the laboratory strain of *Serratia marinorubra*. This could be the

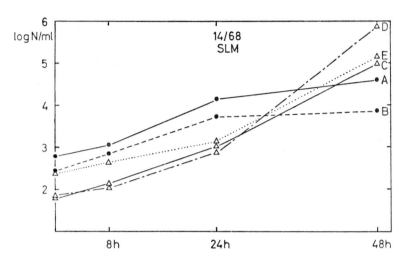

FIG. 9.3. Influence of three different concentrations of the emulsifier Hüls SLM upon a natural population of bacteria present in freshly sampled sea water after different times of storage. A = freshly sampled sea water; B = paper-filtered sea water; C = 1000 mg SLM/litre; D = 100 mg SLM/litre; E = 10 mg SLM/litre.

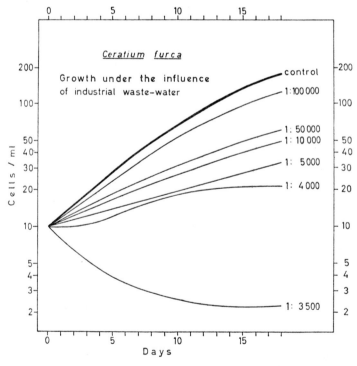

FIG. 9.4. Growth of a unialgal culture (*Ceratium furca*) under the influence of different concentrations of waste-water of a titanium factory.[2]

case also in other organisms, that laboratory strains do not behave like freshly sampled ones.

2. Organic substances can be both toxicants and nutrients for bacteria.

The use of natural populations of bacteria for toxicity testing has several shortcomings:

1. The bacterial populations change in nature continuously in numbers per volume unit, species composition and growth phase.
2. To understand the influence upon a natural population of micro-organisms, the species composition should be monitored. The determination of the different species is a very complicated and time-consuming work which can be done only in special laboratories. However, the response of a natural population of bacteria gives more realistic results for judging the influence of substances upon natural environments.

Experimental Series No. 3 and 4

The following two experimental series were conducted by Kayser.[2,3,4] He used unialgal cultures of the two phytoplanktonic organisms *Ceratium furca* and *Prorocentrum micans*. Both cultures were not bacteria-free. He tested

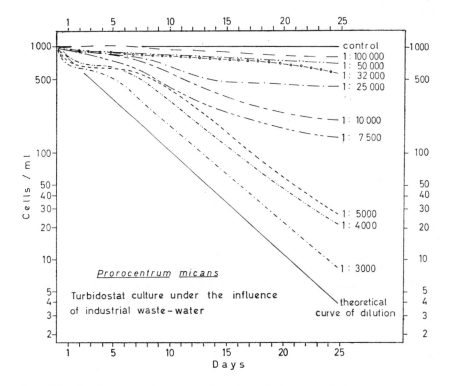

FIG. 9.5. Continuous cultures of the phytoplankton organism *Prorocentrum micans* under the influence of waste-water of a titanium factory[2] (for explanation, see text).

the influence of different dilutions of industrial waste-water of a titanium factory (which consisted mainly of a mixture of diluted sulphuric acid and iron sulphate) upon growth rates. The results are compiled for *Ceratium furca* in Fig. 9.4. During 18 days the numbers of plankton cells in the culture without added waste-water increased from 10 per ml to 200 per ml. Already a dilution of the waste of 1:100 000 resulted in a slower multiplication of the cells. At a dilution of 1:4000 there was only a doubling of the original number of cells within 18 days. This influence would not have been detected using the LD_{50} method, although the toxicant has a strong influence which reduces the gain to 10% of the unpolluted control. Phytoplankton are primary producers, and because these organisms are the first members of the food chain in the oceans they are the basis for all life in the sea. This great ecological significance, together with the possibility of handling high numbers in relatively small experimental set-ups, makes them ideal for testing purposes.

Whilst *Ceratium furca* was used in a batch culture, the following experimental series was conducted using the continuous-culture method. The phytoplankton organism *Prorocentrum micans* served as the test organism. The toxicant was the same as used in the previous experiment. The results are shown in Fig. 9.5. A culture density of 1000 cells per ml was used. In the control experiment without waste-water as much medium was added to keep the number of cells constant. Adding the same amount of medium with different dilutions of the toxicant, the numbers of cells of the cultures decreased because the length of the generation time was extended. Also, this experiment with *Prorocentrum micans* shows that the detrimental effects of toxic substances start far below a concentration which would be detected with the LD_{50} method.

EXPERIMENTS USING FERTILISED EGGS AND LARVAE OF THE HERRING (*Clupea harengus*)

Experimental Series No. 5

Red mud ('Rotschlamm') is a waste product of aluminium smelting. Because it is proposed to dump millions of tons of this material per year in the North Sea, teamwork by several German institutes, including the Biologische Anstalt Helgoland, was conducted.

An experimental dumping of 15 000 tons was made and the influence upon different groups of organisms monitored. Besides this, laboratory experiments about the influence upon different members of the food chain were conducted. Paffenhöfer,[6] for example, investigated the influence of red mud upon *Calanus helgolandicus*. Because these organisms ingested large amounts of the red mud, the growth was retarded, there was a higher mortality and also the reproduction rate was reduced. *Calanus* plays a very important role as food for marine fish.

Rosenthal[8] investigated the influence of different concentrations of red mud upon the heart-beat frequency of embryos, the time the fertilised eggs needed to hatch, and the larval length of the freshly hatched herring *Clupea harengus*. Some of these results are shown in Fig. 9.6. Compared with the

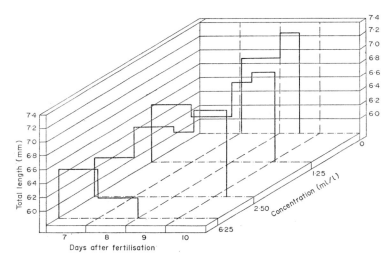

FIG. 9.6. Hatching of herring larvae under the influence of 'Rotschlamm' (red mud). Plotted is the length of freshly hatched larvae against hatching time and concentration of red mud in ml per litre of sea water.[8]

control, addition of red mud resulted in an earlier hatching of the larvae, they were smaller than normal and a high percentage of them showed deformations. The mortality of the embryos shows a strong increase with increasing concentrations of red mud. At a concentration of 1:1000 the mortality was 49·5%, compared with 10·6% of the control experiment. Further experiments about the influence of red mud upon plankton algae and different animals were conducted by Kayser, Apelt, Klöckner and Bulnheim and are summarised in the report of the Biologische Anstalt Helgoland for the year 1971.[1]

Adult fishes are usually much more resistant to toxicants than juvenile life stages and it is conceivable that a population could be wiped out or strongly reduced owing to a low-level poisoning of spawning grounds at the time of the growth of fertilised eggs and the hatching of the larvae.

Experimental Series No. 6

In this series Rosenthal and Gunkel[7] investigated the influence of a mixture of Iraq crude oil with an emulsifier (Moltoklar) mixed in the proportion 4:1 upon herring larvae having a length of 20–26 mm. In control experiments, where the surface of the sea water was covered with a thick layer of crude oil, within 4 days there was no harmful influence noticeable upon the herring larvae. However, within 2 hours at a concentration of 50 mg emulsifier and 200 mg of oil per litre, all larvae died. Definite sublethal influences upon the larvae could be noticed even at a concentration as low as 50 μg after 2 days. The results of one of these series are shown in Fig. 9.7. In these experiments there was a subjective factor, especially in judging the slight influences of activity changes, but it was possible to show that a concentration of several orders of magnitude lower than the one which killed the larvae still exhibited effects which would not have been noticed using the LD_{50} method.

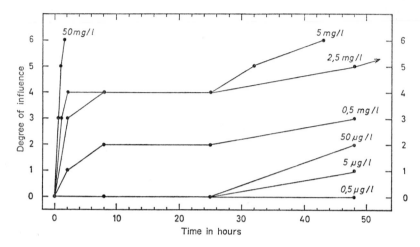

FIG. 9.7. Influence of a mixture of Iraq crude oil with an emulsifier upon herring larvae (20–26 mm length) in dependence of the emulsifier and time. 0 = no influence; 1–3 = small to moderate influence upon activity; 4 = strong influence upon activity, no food uptake; 5 = 50% inactive but alive; 6 = more than 50% dead.[7]

ACKNOWLEDGEMENTS

I am greatly indebted to Mr E. B. Cowell, British Petroleum Co. Ltd, London, for critically reading and improving the manuscript, and to Dr Kayser, Biologische Anstalt Helgoland, for stimulating discussions. Many thanks are due to Dr Bock, Chemische Werke Hüls, Marl, for supplying the emulsifiers.

REFERENCES

1. Biologische Anstalt Helgoland, *Jahresbericht 1971*, Biologische Anstalt Helgoland, Hamburg, 1971, 94 pp.
2. Kayser, H., Züchtungsexperimente an zwei maritimen Flagellaten (*Dinophyta*) und ihre Anwendung im toxikologischen Abwassertest, *Helgoländer Wiss. Meeresunters.*, **19**, 21–44 (1969).
3. Kayser, H., Experimental-ecological investigations on *Phaeocystis poucheti* (Haptophyceae): cultivation and waste water test, *Helgoländer Wiss. Meeresunters.*, **20**, 195–212 (1970).
4. Kayser, H., 'Pollution of the North Sea and Rearing Experiments on Marine Phytoflagellates as an Indication of Resultant Toxicity', in *Advances of Water Pollution Research* (Proc. 5th Int. Conf., San Francisco and Hawaii, 1971), Pergamon Press, Oxford, Vol. 2, III-2/1-7.
5. National Academy of Sciences, *Marine Environmental Quality: Suggested Research Programs for Understanding Man's Effects on the Oceans*, Report of a special study held under the auspices of the Ocean Science Committee of the NAS-NRC Ocean Affairs Board, National Science Foundation, Washington DC, 9–13 Aug. 1971, 107 pp.
6. Paffenhöfer, G.-A., Einfluss von 'Rotschlamm' auf Mortalität und Körpergewicht von Jungstadien des marinen planktischen Copepoden *Calanus helgolandicus*, *Naturwissenschaften*, **58**, 625 (1971).

7. Rosenthal, H., and Gunkel, W., Wirkungen von Rohöl-Emulgatorgemischen auf marine Fischbrut und deren Nährtiere, *Helgoländer Wiss. Meeresunters.*, **16**, 315–20 (1967).
8. Rosenthal, H., Wirkungen von Rotschlamm auf Embryonen und Larven des Herings *Clupea harengus*, *Helgoländer Wiss. Meeresunters.*, **22**, 366–76 (1971).
9. U.S. Department of the Interior, 'Interim Toxicity Procedures' (manuscript), U.S. Department of the Interior, Federal Water Pollution Control Administration, Washington DC, 1969, 10 pp.

Discussion

Mr L. R. Beynon: We are fortunate to have with us Capt. Roland Engdahl, Head of the Swedish Coastguard Service. He is not a biologist, and does not wish to contribute to the Workshop on the subject of toxicity testing. Nevertheless, I feel that it would be valuable to permit Capt. Engdahl to say a few words on oil pollution clean-up, to help set the scene for the remainder of the proceedings.

Capt. R. Engdahl (Swedish Coastguard Service, Stockholm, Sweden): I am in charge of the Swedish operations for dealing with oil spillage at sea which are mainly based on the coastguard organisation. We have some 100 coastguard cutters and other ships as well as 20 specialised pollution treatment vessels. These are located at various places around the coast and the readiness for dealing with oil spills at sea is integrated with the day-to-day work that is performed by the coastguard. This includes surveying the coastal waters and the high seas, such as the Baltic and the western approaches, and also many other activities.

Our philosophy for dealing with oil spills is that you should try to get spilt oil out of the water because, when you have got it out of the water, it will not do any more harm to the marine environment. But as all who have studied this problem or come into contact with it will know, it is much easier to say 'Get the oil out of the water' than to do it. Techniques for containing and picking up oil out of the water have been developing rather well in the last few years, but there is still a lot of work to be done in this field. Even when we have the necessary equipment available, however, there will be occasions when even the most sophisticated techniques will not help very much because, for example, there is very rough weather or the oil spill has had time to spread over a wide area. For such reasons it is necessary, especially at present, to have alternative ways of dealing with oil spills. Sometimes we use absorbents, but, as we all know, they also have great limitations, especially when dealing with large oil spills. Even if you can absorb the oil, you are faced with the problem of recovering the oil-soaked material. Burning oil is another possibility and the Finnish organisations have had some success in developing techniques for burning oil on the sea and in ice conditions. If you have great quantities of spilt oil on water, however, burning it may turn the water pollution into air pollution, and careful consideration has to be given as to whether this is acceptable. There is another method for dealing with oil pollution—sinking the oil. It is a method that we definitely do not approve of in Sweden because no one really knows what happens to the oil that has been sunk. The fifth method is that which has been discussed here today and will be discussed further in this Workshop, that is, dispersing the oil.

We have paid a lot of attention to developing techniques and tactics for using dispersants and we are well aware of all the potential ecological problems that are involved with dispersant use and misuse. We are particularly happy that so many skilful people are working in this field to provide us, who have to work at the front, with the information on how to act and indeed when to act. I should like, however, to add one piece of information concerning the use of dispersants. An

87

effect that is very seldom pointed out is that even if you do not succeed in dispersing the oil into small droplets, which is the main object in using dispersants, you will have one very essential advantage. We have found with Corexit 7664, and I believe that it is the same with BP 1100X, that treated but undispersed oil which comes into shallow waters and into the beaches will not coat the beach or stick to the flora and fauna in the littoral zone. When the water later moves out, whether by tide, wind or current, the oil will also tend to move out and be dispersed in the way that was originally intended. In this connection, it is of interest just to speculate for a moment on what could have been done in the West Falmouth case where a lot of diesel oil drifted into the narrow and shallow waters of West Falmouth in the USA, coated and mixed with the bottom sediments and killed most of the flora and fauna in the area. What would have been the result if it had been possible to treat this diesel oil so that the oil droplets would have been coated with dispersant? Would it have had the same effect? I have discussed this point with several people and we have come to the conclusion that most probably there would have been much less harm caused, perhaps no harm at all.

Mr L. R. Beynon: We have come to the end of the formal presentations. I am hoping that all the biologists from other laboratories will speak for a short while on the work that they are doing. But firstly, in view of the interest already expressed in the research of Dr B. Ganning, I am going to invite him to give us some further details of the activities of the Askö Research Laboratory in Sweden.

Dr B. Ganning: First of all I want to tell you something about the specific problems of the Baltic, which is the largest brackish water area in the world. The salinities are low compared with those in the North Sea, which are about 30–35‰. In the Southern Baltic we have 10‰ and in the Northern Baltic 2‰ salinity. As a result, the Baltic animals and vegetation differ greatly, both in species and number, from those found on North Sea coasts. On the Swedish west coast there are something between 1100 and 1500 different macroscopic species, whereas just outside Stockholm, between Stockholm and Helsinki in Finland, there are only 52 macroscopic animal species. Therefore, in theory at least, it is possible to test all these species in toxicity studies. Another problem with the Baltic is that the water is retained for a very long time. The mean time for water to stay in the Central Baltic is 30 years. In the Kattegat we consider that the water is changed four times a year. Since 30 years is the mean retention time, surface water may stay in the Baltic for 1–5 years whereas the bottom water may stay for perhaps 100–300 years. Another problem is related to the Baltic fjords, which have very shallow entrances with a maximum depth of 80 metres. The maximum sea depth is 459 metres and there is a large area of 250 metres depth. The water in the top 60 metres is clearly and absolutely separated from the deeper waters because of the halocline, and there is very little mixing between the surface and the bottom water. The next problem is that there is virtually no tide in the Baltic; the yearly range is about 1·5 metres caused by changes in barometric pressure, winds, etc. The water is very cold, with a mean annual temperature between 5° and 8°C and ice cover during a considerable part of the year, which again creates great problems with respect to oil pollution. What I am now going to tell you about is my basic research work at the Askö Laboratory of the University of Stockholm.

I share the opinions of the biologists from the Orielton Field Centre in that I am not a great believer in the value of LD_{50} tests as described by other people at this Workshop. Let me give one example illustrating why it is very dangerous to measure only LD_{50} values. One of the 52 animal species in the Baltic is a crustacean. I tested it in the common way with a highly toxic dispersant which was used in 1969 in the archipelago of Stockholm, and found that the LD_{50} value for

96 hours was about 100 ppm. However, I wanted to have some measure of sub-lethal effects and to do this I made use of the fact that the male carried the female for between one day and a fortnight before copulation. I observed when the male lost his interest in the female, and that occurred at 10 ppm. The fact that at 10 ppm no copulation can occur is of much more interest to biologists than the fact that the adult animals die at 100 ppm. One problem with a sublethal study is that it is very difficult to have a statistical quantification of what happens, and perhaps the best way is to count how many couples have a 'divorce'.

The females carry their young in a brood pouch on the stomach and, at con-centrations as low as 0·1 ppm, the young are spontaneously aborted and die very quickly.

I have also used another sublethal test which may be of interest. I wanted to study the primary production of the marine community in water polluted by oil and dispersed oil. In my tests I used *Fucus vesiculosus*, which is one of the ten large algae entering the Baltic; we also have about ten microscopic algae. Now the most common oil transported in the Baltic is No. 4 fuel oil, and I compared the effects on *Fucus vesiculosus* of various concentrations of this oil with control samples. I also studied two dispersants, BP 1100 and Corexit 7664.

I have test tanks containing a control, oil at 1000 ppm, and oil dispersed in various concentrations of BP 1100 and Corexit 7664. Oil and dispersant were used in a 1:1 ratio. Although the work is still continuing, I have already had some interesting results.

In the test tank containing floating oil the water eventually turned brown, showing that the *Fucus* was losing pigment. I also measured the oxygen concentra-tion and the diel O_2 curve over 24 hours. The latter is a measure of the primary production and the community restoration during the 24-hour period and is a fairly sensitive way of studying the effects of oil and dispersed oil. At low pollution concentrations you have an autotrophic system with a production rate greater than the respiration, but at high concentrations respiration greatly exceeds the primary production and inhibition of photosynthesis takes place. This effect is particularly marked for Corexit 7664 when, at only 560 ppm, there is improper primary production by these algae, and at 5600 ppm there is almost no primary production at all but very great respiration.

These are some of the ways I have studied oil pollution and tried to find a sensitive way of determining what happens to very sensitive Baltic ecological systems. You must remember, however, that it is impossible to translate the figures or results obtained from British investigations to the Baltic situation. In the Baltic we have some fresh-water animals and some marine animals, a non-tidal region and low water temperatures, which is a very different situation from that which pertains elsewhere in Europe.

In conclusion, I want to quote some figures reported at a Swedish–Soviet Union symposium, although I know that Swedish researchers do not agree with the data. The Russians stated that, owing to the very slow water exchange in the Baltic, its oil content is between 1 and 3 mg/litre. In Sweden we feel that the true value may be one-tenth or one-hundredth of that figure. If the Russian calculation is right there are about 2 million tons of oil in the Baltic, so that the Baltic is soon going to be a fine oil well. But even if their results are wrong, it is important to remember that what is let into the Baltic stays a very long time.

Mrs C. van der Wielen (Institute of Hygiene and Epidemiology, Brussels, Belgium): A year ago, our Institute was entrusted with studying the toxicity and biodegrada-tion of dispersants. The purpose was mainly to make a selection from numerous commercially available products. The experimental part of this work has scarcely started and therefore I shall draw attention only to a few points which have particularly interested us.

On what criterion should we give preference to one product rather than to another? A better knowledge of the mode of action of those products would enable us to choose them and to use them better. At the moment we generally choose them only on the basis of a measure of 'mortality'. I should like to add to this first criterion the idea of 'speed of action'. The determination of the TL_{50} after 24 or 48 hours, or the growth curve of algae, exposes a level of toxicity after a time of 24 hours or more, but this figure does not tell us whether a dispersant has had progressive effects or if, on the contrary, it has had more or less immediate effect. This seems to me, however, a useful guide towards a choice of dispersant.

At our Institute we tested about 20 dispersants with the larvae of the barnacle *Elminius modestus*. The simplicity of the test allows one to follow the progressive deterioration of a population of larvae. Some results obtained are given in Table I. We noted TL_{50} values after 5 minutes, 15 minutes, 1 hour and so on until 24 hours and, for example, as seen from Table I, the dispersants E_{10} and E_{13} have the same TL_{50} value after 24 hours, equal to 0·75 ppm. However, their speeds of action are

TABLE I

Toxicity of Commercial Dispersants to Larvae of *Elminius modestus*

	TL_{50} 5 mins	TL_{50} 15 mins	TL_{50} 1 hr	TL_{50} 24 hrs
Water-soluble products				
E_2	100 000 ppm	100 000 ppm	37 500 ppm	9 500 ppm
E_3	37 500	37 500	25 000	2·5
E_1	25 000	12 500	4 800	4 800
E_4	6 300	1 200	1 200	690
E_5	630	40	30	15
Petroleum hydrocarbon-based dispersants				
E_7	10 000	10 000	10 000	10 000
E_{13}	125	16	6	0·75
E_6	48	6	3	3
E_{11}	48	12	6	3
E_8	48	16	6	0·75
E_{16}	32	12	6	2
E_{12}	24	16	3	0·75
E_{22}	24	12	4	2
E_{17}	16	5	2	1
E_{18}	12	6	3	1
E_{21}	12	6	3	1·5
E_{20}	8	6	3	1·5
E_{23}	8	5	2	2
E_{19}	6	3	1	1
E_{14}	6	3	2	1·5
E_{10}	6	2	0·75	0·75
E_9	6	3	3	3
E_{15}	4	2	1	0·5

quite different; for E_{13} after 5 minutes, the TL_{50} value is 125 ppm; after 15 minutes, 16 ppm; after 1 hour, 6 ppm. For E_{10} the TL_{50} is already 6 ppm after 5 minutes, 2 ppm after 15 minutes, and 0·75 ppm after 1 hour. I do not know what is the speed of the dilution of the dispersants. As regards the experimental conditions, we consider of particular importance agitation, temperature, sea-water composition and the nature of the polluting oil.

In the case of an agitated sea, the products will be better emulsified and will then have more chance of being in close contact with the organisms. I believe it is useful to test the most unfavourable condition for the fauna and flora. We have therefore evolved a system of tables of agitation for all cultures of algae.

What is the biological effect of a film of hydrocarbon on the water surface and of treatment by dispersants during the cold period of the year? Most tests I have seen have been conducted around 18°C. The physical chemistry of the products is not indifferent to temperature, and who knows if classification would not be very much modified at other temperatures? In addition, at low temperatures one has to utilise much more dispersant.

In relation to the nature of sea water used, we have made a compromise between natural conditions and standardisation. So, for five replications, three samples will be synthetic sea water, and two North Sea water. It is only when there is dissimilarity that we shall carry out multiple experiments.

Among the pollutants to consider, there are weathered products. In the preparation phase we have checked the quantity of inorganic salts necessary for biodegradation. Indeed, the threshold of toxicity being sometimes relatively high, we must start from relatively large concentrations of toxic substances in order to be able to follow the changes in toxicity with time. In this case (for example, solution of 1% or 0·1% in dispersant) the quantities of inorganic salts proposed by manuals such as the OECD manuals (around 10 ppm) were insufficient and were a limiting factor.

The results presented in graphical form are expressed as percentage of limitations in relation to the results obtained with the highest concentration of nitrogen.

Because of our geographical situation, far from the sea, we can obtain stocks of marine organisms only periodically. At the moment we carry out continuously three types of test:

1. Influence on autorespiration.
2. Determination of the TL_{50} value for the fish *Lebistes reticulatus*.
3. The growth of phytoplankton.

From the first test it was observed whether, by addition of a pollutant, there is inhibition or not of the degradation of a standard peptone solution indicated by BOD value. Oxygen consumption measurements are made on an hourly basis, from which percentage inhibition is calculated. Experiments by other workers suggest, however, that this test is relatively insensitive towards dispersants.

In spite of its great resistance, we have selected the fish *Lebistes reticulatus* for the following reasons:

1. One can easily maintain and reproduce them in the laboratory. This allows for low cost and use of a statistically significant number of organisms.
2. Because of their small size, the volume of water necessary for the experiments may be significantly reduced.
3. One can easily acclimatise them to saline conditions.

We practised static bioassay as described in standard methods. For better standardisation in this case, we needed to control the quantity of pollutant introduced, and we also determined the COD value of the solutions. More recently we have determined TLC values, which seem more precise but raise some problems because of the lipophilic nature of some of the products.

Before and after degradation we measured the amount of chemically oxidisable matter, and thus obtained COD values for several samples of Corexit 7664 with different concentrations of nitrogen (see Table II). For example, for one level of nitrogen the COD value dropped from 7380 to 5320 mg/litre after degradation so that 27% of the chemically oxidisable matter had disappeared. For the TLC measure we obtained, for the same concentration, 1985 mg/litre. Lastly, we made measurements of the relatively stable part of some emulsions. We mixed the emulsions and let them stand for 15 minutes. We determined the COD of the mixed emulsions and of the decanted parts and obtained for 50 g/litre solution:

(a) for Corexit 7664, 35 000 mg/litre before and after decantation (it is soluble);
(b) for Finasol OSR, 32 000 before and 15 200 mg/litre after decantation;
(c) for Elimax, 31 200 and 22 400 mg/litre after decantation.

We could thus ensure from test to test that we worked on the same amount of product.

TABLE II

**Amount of Chemical Oxygen Demand of Corexit 7664
Before and After Biodegradation**

	mg/litre before	mg/litre after
COD 1% Corexit	7140	6920
	7160	6840
	7220	5920
	7220	5820
	7380	5320 → 27·8% disappeared
COD 0·1% Corexit	738	676
	744	562
	752	524
	752	532
	744	540 → 27·4% disappeared
TLC of 1% Corexit	1985	
0·32%	642	
0·1%	195	

Dr A. R. D. Stebbing (Institute for Marine Environmental Research (IMER), Plymouth): One of our projects is the development of new techniques for bioassay of pollutants. When we started, about a year ago, it appeared that the most important requirement was for bioassay techniques with greater sensitivity. It seemed that what we really want to know from the point of view of the environment is the lowest concentration of a pollutant at which there is a measurable deleterious effect upon test organisms. Our methods are therefore intended solely for the study of the effects of sublethal concentrations of pollutants. Currently we are trying out two possible methods.

The first method is one using the settlement performance of *Spirorbis* larvae as it might relate to varying concentrations of different pollutants. As you probably know, these larvae have a complex pattern of behaviour before finally settling and metamorphosing which one would expect would be sensitive to pollutants. The relationship between the numbers available to settle and the numbers that actually succeed in doing so seems to be a good measure of performance. A number of people have suggested that the larval stage of marine organisms is the most vulnerable, but the only work that I have found in the literature is concerned solely with mortalities of larvae in relation to pollutants and not with their ability to settle and metamorphose into adults. It is possible that much lower concentrations than would kill them could so influence their settlement behaviour that the effect is the same as a lethal concentration. This would be similar to the 'divorcee isopods' that Dr Ganning discussed; presumably after divorce they did not reproduce. Our work with *Spirorbis* is at an early stage; the first season we have used to devise a completely artificial substratum for the larvae to settle on. It was essential to do this because discs of their natural substratum varied so much in their attractiveness to the larvae, in that they might prefer discs from one fucoid frond to discs from another and they even showed very marked preferences for different parts of the same frond. In fact in *Laminaria* I have managed to show that there is an ecological advantage in the preferences that larvae of some bryozoans and serpulids show for the youngest parts of the *Laminaria* fronds. With artificial discs, which we have managed to develop in this first season, we can ensure that they are equally attractive to the larvae so that replicates are comparable. However, this approach has obvious disadvantages; *Spirorbis* liberates larvae for only six months each year and the adults have to be collected from the shore where they have been suffering an unknown spectrum of pollutants. So we have also taken another approach, although we are going to continue with the *Spirorbis* larvae work because I believe that it is going to be rewarding.

The second approach avoids many disadvantages of the first. We are trying to use the growth rates of cultured hydroid colonies as a bioassay technique. These can be cultured indefinitely in the laboratory using *Artemia* larvae as food, and what we are hoping is that the growth rates of the colonies will reflect the concentrations of the pollutant to which they are subjected. One of the major advantages of using hydroids is that one can avoid genetic variations in growth rate by using hydrants from the same colony or clone in each experiment. This work too is at an early stage.

We started by using *Hydractinia* but we found that their growth rate was rather slow, besides which they apparently need a specific bacterial film on which to grow. We are going to try another species (*Podocarynie*) and we know from work in the United States that it will be much easier to work with; in fact we shall be able to use this technique in its entirety. One other thing that I should point out before I finish is that all the work we are doing with both *Spirorbis* and *Hydractinia* we are adapting for use in artificial sea water, so that we have a stable base line in time.

So our aims are clear. We are searching for robust yet sensitive bioassay techniques which we can use at any time of the year, and with which we hope we shall get consistent results. However, we have a long way to go before we begin to achieve these ends.

Mr A. D. McIntyre (Department of Agriculture & Fisheries for Scotland, Marine Laboratory, Aberdeen, Scotland): I should say first perhaps that the Marine Laboratory in Aberdeen has, at this time, three active interests in oil. First of all, we may be called in if there are any oil spills or disasters. Secondly, we are concerned in association with the Ministry of Agriculture, Fisheries and Food in a study around the British coast of the background levels of oil in the sea. Thirdly, and this is what I want to refer very briefly to today, we are interested in the long-term effect of oil at low concentrations in marine animals; not just in the toxic effects in terms of death but also in the effects the oil may have in contaminating the flesh. In this context we have just started some experimental work in co-operation with the Torrey Research Station in Aberdeen.

In this work we are using codling (*Gadus callarias*), that is, fish about 30 cm in length, and also scallops (*Pecten* spp.). The animals are kept in large fibre-glass outdoor tanks, the dimensions of which are 12 ft by 6 ft by 4 ft high. These experiments are intended to run for about a year and are in two parts. In the first phase, which lasts for six months, we are keeping the fish and the shellfish in oily water, and we are feeding the fish with oil and taking off fish approximately every six weeks for analysis. After six months the fish are fed only on clean food and the fish and shellfish are kept in clean water, and we hope then to follow the rundown of any contamination of the flesh. I am afraid that it is not possible to give any results yet because this work has only just started. But I thought that this long-term type of experiment might be of interest and I can give further details of the techniques we are using if anyone is sufficiently interested. We hope soon to be able to report some results of this work. I think that is all I want to say in this connection, but perhaps Dr Johnson, also of Aberdeen, may want to add something later about some of the other aspects introduced.

Dr R. A. A. Blackman (Ministry of Agriculture, Fisheries and Food, Fisheries Laboratory, Burnham-on-Crouch, England): Capt. Engdahl mentioned that Swedes frowned very much on the method developed by the Dutch of dealing with a large oil spill by sinking it. I have a few observations on the effects of sunken oil that we have made in the laboratory. These are not strictly the results of toxicity tests but they do have some relevance to long-term tests with oil, and they may be of use to anybody setting up such work. I should perhaps explain that this method of dealing with oil is to spray the slick with a slurry of sand, sea water and a wetting agent, usually a tallow amine acetate, which causes the oil to adhere to the sand. The combined density then rises above that of sea water and the masses are sunk to the bottom.

The first thing we found was that brown shrimp (*Crangon crangon*), edible crab (*Cancer pagurus*), the shore crab (*Carcinus maenas*) and the lobster (*Homarus vulgaris*) will all eat crude oil sunk in this way. In the case of the brown shrimp the oil remains as compact masses in the foregut and in the gastric mill, which is the grinding apparatus for food, until the next moult. We also found that brown shrimps that had been lightly oiled externally with untreated crude oil will clean the oil from their forelimbs and from their antennae if they are placed in clean water, and they ingest part of the oil in the process. This could be particularly relevant to the method of toxicity testing by dipping the animal in oil, etc., and then placing it in clean conditions. If the ingested untreated oil behaves in the same way as the ingested sunken oil, it will remain until the next moult and any longer-term effect will be at least partly due to internal absorption of toxins as well as to external absorption through the carapace. We have carried out some experiments with brown shrimps to follow the effects of sunken oil. I will admit that they started out as toxicity tests of a kind. The first experiments were carried out in February–March at a tank temperature of 9·5°C and using sunken topped crude oil. There was no significant mortality over 40 days in contact with the oil, but the feeding rate on normal foods supplied was reduced in shrimps in contact with the oil. The exposure to oil solubles did not affect the feeding rate. We carried out another experiment in May–June with tank temperatures averaging 15·5°C using sunken fresh crude oil. There was then a significant mortality over 40 days. The feeding rate of normal foods was completely unaffected. In both experiments more unfed than fed animals ate sunken oil, but fewer animals ate fresh crude oil and smaller quantities of fresh crude were eaten than the sunken crude oil, and in both experiments the moulting rate of the shrimps was un-affected. It was unclear from these experiments whether the shrimps were responding to amphotenic substances in the oil itself, perhaps to the tallow amine acetate that was used to sink the oil, or whether the normal process used to discriminate food was being interfered with. We have therefore carried out some supporting experiments using normal foods which contained 1% of the sunken fresh crude oil. This was obtained by centrifuging the sunken masses. In clean conditions the feeding rate was reduced on this oily food. In contact with sunken oil the feeding rate on clean food was unaffected, as reported before, but after exposure to contact with this oil for 40 days the feeding rate on oily food, when replaced in clean conditions, was also unaffected. So the preliminary conclusion that one can come to is that the sunken oil is itself distasteful, but that enforced contact with the sunken fresh crude oil interferes with the food discriminatory process. We hope to follow this up further.

Dr B. Ganning: I have a question on oil-sinking. When you sink oil, what does it look like in your aquarium, on the bottom? Do you have a sheet of oil on the bottom, or do you have droplets, or do you have lumps of oil covering the bottom?

Dr R. A. A. Blackman: If you sink oil so that there is not sufficient concentration to form a complete layer on the bottom, then it forms lumps of something like 1 cm diameter, but this is in a laboratory; we are only sinking through a depth of water to perhaps 30 cm. The oil may be more broken up than that in deeper water.

Mr L. R. Beynon: In certain field experiments, using not only sand as a sinking agent but also other materials, all sorts of sizes of lumps have been found fairly well dispersed on the bottom. The sizes ranged from 1 cm to perhaps 10 cm.

Mr H. J. Marcinowski (Stichting CONCAWE, The Hague, Netherlands): In the experiments mentioned by Capt. Engdahl which involved 100 tons of Arabian crude, it was found that some days after the oil was sunk it was covered with a sand layer, and the oil/sand mixture was more or less a 'sandy cream'. Later, through a rubbing effect of sand movement, over a period of about four weeks, some oil came to the surface. The quantities of oil surfacing steadily decreased, and the largest droplet sizes rising from a depth of 22–25 metres was less than 0.25 cm^3.

Mr L. R. Beynon: This sort of field experiment is very difficult, but some experience has been gained from actual pollution incidents. For example, after the *Torrey Canyon* incident the French authorities tried to sink a lot of chocolate mousse using their 'craie de Champagne', and the oil certainly went down below the surface. Divers went down some time later and could not find any oil on the bottom. It was suggested that this was because it had all degraded, but this seems unlikely. I know of some trials undertaken by the Department of Trade and Industry in the UK using the French sinking agent. Divers went down immediately afterwards but could not find anything at all. It had gone somewhere, but presumably bottom currents and sand movements had dispersed it.

10
A Critical Examination of Present Practice

E. B. COWELL

(British Petroleum Co. Ltd, England)

This Workshop[1] was convened to examine the approaches and procedures that are used by various laboratories in Europe. If the Workshop is to achieve its desired objective, then it behoves us to make a critical and honest evaluation of those practices with which we are all involved, so that future developments represent a real improvement rather than a habitual continuation of tradition. We are all bound to have our preferences for the approaches that we adopt, the reasons for which may be personal rather than scientific.

From the lead papers already presented[1] and from an examination of toxicity-testing literature it is immediately clear that a great deal of confusion surrounds the whole subject. Much of this confusion can be blamed upon ourselves as biologists, partly for using an excess of cryptic jargon in communicating (or failing to communicate) what we are attempting to do, and partly, I suspect, because we are not always sure why we do toxicity-testing work in the first place.

While remembering that this Workshop was specifically convened to examine the problems of toxicity testing of oils and dispersants, nevertheless the papers presented generate a number of fundamental questions about toxicity tests in general. The first of these is: Why do we do toxicity tests? While superficially this may seem to be a naïve question, I put it to you that this question is not asked often enough, nor critically enough, and worse, often not until the work has been started, experiments designed, equipment purchased and money spent. Sometimes one suspects that toxicity testing is used simply as a rapid means of getting large numbers of publications into the scientific literature. All those biologists attending this symposium, including myself, are presently or have recently been involved in toxicity-testing work. That I pose the question 'Why do we do these tests?' is therefore not an implied criticism of any one individual or organisation, but rather a plea for a critical objective examination of the whole subject, our part in it, and the part played (whether reasonable or unreasonable) of those who demand toxicity tests from us.

Personally I believe that laboratory bench toxicity tests have a very limited but not unimportant usefulness in reaching the ultimate aim of ecologists involved in environmental protection—this aim being, of course, to enable assessments to be made of the present or possible future impact of pollutant materials upon ecosystems. Final judgement must always be made on the ecosystem response measured in terms of changes in populations, community interactions, diversity, stability, production and other parameters.

If we assume for the moment that we know why we do toxicity tests, the next questions to be asked are: How should our tests be conducted? How should data be presented? What limitations should govern our interpretation of the data collected?

There may be many reasons for requiring toxicity data on oils, dispersants and oil/dispersant mixtures. The type of test used will largely be governed by specific questions that we ask before setting up our laboratory procedures. From the papers presented at this Workshop, it is clear that toxicity tests fall into two main categories:

1. Tests to produce comparative rankings of toxic materials in some standardised form.
2. Tests from which ecological predictions can be made.

These two types, while often complementary, cannot and do not fulfil both requirements. Comparative rankings are of limited value, but are required by governments, oil companies, dispersant manufacturers and others to assess the relative value or risks involved in the case of various materials against other materials which have already been experienced. The latter, of course, often provide convenient reference materials.[2,3] The information obtained tells us nothing on its own except the order of toxicities obtained under laboratory conditions.

Provided this important limitation is accepted, then the next task to be undertaken is to examine the types of laboratory procedure available. A great deal of discussion would be of value at this point, but I propose only to present some of the problems that require answering and which I hope will be discussed.

Two papers in this Workshop suggest that toxicity ranking does not differ significantly with different test organisms.[2,4] The Fisheries Laboratory, Burnham-on-Crouch, and the Field Studies Council Oil Pollution Research Unit do not believe after comprehensive comparisons that the species selected is of major importance.[2,4] Certainly, their data and much of the literature supports this suggestion. If the point is valid then the choice of test organisms can be greatly simplified. It can vary from laboratory to laboratory according to factors such as laboratory convenience, abundance of supply, etc., without significantly changing the type of result obtained. It may well be possible, for example, to reduce the whole procedure to a few hours work on yeast-cell respiration measured in a Gilson respirometer or other modification of a Warburg apparatus.[5] In this way laboratory space required is minimal and the test organism can be stored as a dried powder on a shelf. I believe the approach is now practical and could result in a considerable saving of time, laboratory space and money.

When deciding which species to use, is it justified to select species to include representatives of various types of organism, taxonomic groups, etc?[6,7] I doubt that it is, and would like to explain my reasoning for these doubts. Firstly, the selection of representatives of taxonomic groups, e.g.

arthropod, crustacean, mollusc. There is a voluminous literature which adequately shows that in toxin response no one species can adequately represent even one genus, let alone a group of genera. The Orielton work of Crapp[3] and Ottway[8] shows that enormous variation in toxicity response is found even within the genus *Littorina*, while Cowell and Hainsworth,[9] working on salinity tolerance, found that even within the species *Nucella lapillus* response varied with age and within species while *Littorina saxalitis* response varied according to varying physiological races and local phenotypes.

The same type of argument also applies to trophic level representatives and species chosen to represent different types of feeding mechanism.

Data on one species tell us little or nothing about the toxicity response of another species.[3] It may of course give secondary data on mechanisms of reaction or types of behavioural response,[10] although these are better investigated in separate experiments asking quite different questions. At this level, experiments investigating sublethal concentrations are not strictly toxicity trials. If these complex and difficult questions on choice of test organisms are satisfactorily answered, then secondary questions immediately present themselves. These include: How should test organisms be obtained? Should they be bred in the laboratory or collected from the field?[2,6,7] I suspect that the answers will vary with the laboratory situation.

If test organisms are collected from the field, what factors govern our choice of collecting site? Can natural populations ever be said to be 'homogeneous'?[6] I do not believe that they are, nor that we know or understand what we mean by biological homogeneity.

From our collections, how should we select our test animals? Should we, for example, choose our animals from within a particular size range?[3,6] There are hidden dangers in so doing, for every ecologist knows that animals of equal size may vary widely in age, sexual maturity and reproductive state. Should we therefore, as an improvement on deliberate selection, choose our test animals randomly using standard randomising techniques? I believe that we should, but even if this question is resolved, others follow.

Is it necessary in all cases to have periods of laboratory acclimatisation of test animals?[3,4,6,7] If so, should we really conduct tests at temperatures that are the same within the laboratory at all times of year when temperatures vary in the environment? Should we perhaps vary our laboratory temperature to equate it to that from which our animals were obtained, as did Crapp[3] at Orielton? The physiological state of organisms may be closely dependent upon the temperature levels found in nature, and changing temperatures may impose stress and changes in metabolic processes, reproductive condition, etc. There are possibly unjustified assumptions in believing that acclimatisation periods reduce the effect of stress on test organisms and make tests done at different times of year more comparable. In some instances temperature change is known to increase stress,[7] for example in species which release gametes as a response to temperature change. In many cases comparisons of tests performed at different times of year are invalid except for examining effects of season on response.[3,11]

What is the optimum number of animals for each experimental treatment from which acceptably low standard errors are obtained on the results? Standard methods exist for calculating the required number of experimental organisms,[12] yet we have seen that it is common practice to use arbitrary numbers of 10 or 20[2,3,6,7,8] animals. Our own laboratory experience must tell us that frequently biological variation may be great and that optimal numbers will vary from one species to another or even within one species under certain circumstances such as differing age or size groups or differing season. It borders on to malpractice to advise governments or companies on the basis of toxicity tests conducted with insufficient test organisms with large standard errors, yet reports are often submitted by well-paid consultants to biologically ignorant customers which will not stand up to scientific scrutiny.

If we follow up these comments I believe that we must decide that preliminary uniformity trials will be required to decide optimal test numbers of animals. Tests are sometimes conducted with 10 animals when 50 or 100 may be necessary to have confidence in results.

If we have established the optimum and practical numbers of animals for our tests, we should then decide whether animals should be randomised to 'treatments'. I believe that critical analysis of the problem and the statistical requirements for valid analysis will dictate that we should.[12] Organisms taken out of stock tanks tend to be sorted into groups with very different characteristics. For example, in some cases differences will exist in activity; those animals easiest to catch could be put on one test treatment while all those hardest to catch could finish on another. Toxicity response in the two groups may be very different.

To achieve full statistical validity it is also important to randomise each experimental tank of randomly selected animals to their experimental treatment and hence to its position in the laboratory.[12] Laboratory position can significantly change results in situations where temperature, light and other factors are not specially equated.

Many toxicity tests are conducted without sufficient replication. How many replicates are required? Much will depend, of course, on the accuracy required in the data.[12] Replicates may be required simultaneously or with time intervals,[12] as were those of Crapp investigating seasonal changes in toxicity response.[3,11]

These comments on statistical requirements may seem at first to be unnecessary, yet such care has long been practised by agricultural scientists[12,13] and there is an extensive literature documenting the traps that can be encountered when the statistical validity of experimental design is ignored.[12] The final choice of design, animal number, etc., will in most cases be a compromise between what is theoretically desirable and practically possible. Nevertheless it is essential that the experimenter should be aware of the problems if he is to know just how far compromise can be taken and just where experimental errors can be reduced or become too large to be acceptable.

We have seen that some workers conduct their toxicity tests mostly in winter.[7] Other laboratories work mostly in summer.[4] Some assumptions here warrant critical evaluation since organisms are known to vary their response to toxins in relation to many factors. These include growth factors, metabolic rates, feeding responses, etc., which vary enormously from season to season. None of these complications can conveniently be ignored, yet all too often they are. It is also wrong to assume that toxicity response increases with increase in temperature. Ottway[8] has clearly shown that this is not always the case. Ecological predictions based upon tests concluded always at one time of year may be totally wrong when applied at another season.

There are clearly many other biological complications that should be considered. There will be many that I have not thought of, but in the limited time that I have available I should like to examine some other experimental factors that should be taken into consideration. An examination of the papers presented in this Workshop and of the wide literature will reveal many anomalies. Enormous variations can be found between the approaches of different laboratories.

One example that must be given consideration is that of the aeration of test chambers in both long- and short-term experiments. Some laboratories consider aeration to be necessary[2,3] while others claim that it is unnecessary, or undesirable,[6,15] because it changes the nature of oils under test by speeding up evaporation of light volatile fractions. If we examine the papers of the 'no oxygenation' lobby I believe that their own data often reveal that oxygen deficits can and do occur in non-aerated tanks. When this happens, then oxygen can become a limiting factor masking the pollutant effect or in some cases possibly exaggerating it, depending on the type of experiment or test organism. Dr Spooner[15] reports oxygen problems but does not consider the effect serious, while at Endoume[6] toxicity is said to be reduced by aeration of test chambers. I believe that the evidence to support the need for aeration is overwhelming.

Other factors can also induce stress; these include salinity levels,[9] temperature fluctuations and differences between chambers and variation in lighting conditions in different parts of the laboratory. If we can define ways of reducing experimental variables to within statistically acceptable limits, then we shall have made enormous progress as a result of this Workshop.

Fundamental problems exist in testing oil levels in standard equipment. Hydrocarbon chemists express serious doubts on the probability of success in maintaining constant oil concentrations in experimental chambers owing to adsorption of the oil on to the materials from which apparatus is constructed. Even bigger problems exist in maintaining levels of emulsions and dispersions. The mixing of these in experimental sea water is related to the mixing energy available not only at the time of the preparation of dilutions but also throughout its residence in the test apparatus.

In addition, solutions, emulsions and dispersions of hydrocarbons are notoriously difficult to achieve and monitor. Biologists usually know little about these problems, yet few seek advice from competent hydrocarbon chemists before designing and conducting experiments. If biologists' credibility is to be maintained, the views of people with expertise in hydrocarbon chemistry must not be lightly dismissed. The limitations of this side of toxicity-testing work must be freely acknowledged; it should not, however, prevent work to produce the best answers obtainable, but rather act as a stimulus to improve our laboratory practice.

In the field of co-operation between biologists and chemists many problems could gainfully be examined experimentally. For example, does emulsion particle size materially affect toxicity? I believe that it does and that elucidation of the effects would explain many of the discrepancies found in toxicity data produced at different laboratories.[2] I hope that in the ensuing discussion some of the chemists present will provide suggestions for future consideration.

The type of equipment chosen by various laboratories presents a number of problems requiring resolution. Serious criticisms can be levelled at both static tank methods[2,5,8] and continuous-flow equipment.[7] While I appreciate the theoretical advantages of continuous-flow equipment when testing water-soluble materials, I remain unconvinced that we have yet resolved the special problems of oil and dispersant materials in this equipment. I shall list some of my objections as a basis for discussion.

Firstly, I am not satisfied that the 'solutions' and dispersions put into the equipment at one end are those to which test animals are exposed within the test chambers. The materials from which equipment is constructed may be attacked by the test material, hydrocarbons being differentially absorbed onto glass surfaces and absorbed by plastics and rubber, etc.

Secondly, tubing other than nylon presents special problems. Many oils and solvent-based dispersants attack the plasticisers in plastic tubing, possibly releasing them into test tanks. Some of these plasticisers are themselves highly toxic. Many accounts of tests in the literature admit to problems arising from hydrocarbon damage to components of test equipment, including tubing joints and pump glands. I have experienced such problems myself and that problems exist is sufficient warning.

Thirdly, water exchange rates in continuous-flow equipment may be poor owing to friction effects, current eddies and inefficient mixing. If this is the case, then animals may be able to select sections of test chambers with low concentrations. Under some conditions, build-up of toxic materials is also possible.

Dr Gunkel[17] has already mentioned some of the problems that can arise from bacterial infections in test tanks and during storage: not only may pathogens be present, but also bacteria which attack oils themselves. These organisms may also grow in vessels used to store test materials and standard dilutions. Microbiological problems in oil storage are well known to the oil industry and there is an extensive literature on this subject. Bacterial hazards in biological experiments certainly exist. We experienced some during my time with the Oil Pollution Research Unit at Orielton.

Field trials for examining ecological effects of pollutants are clearly enormously difficult to conduct yet provide the only ultimate approach from which realistic appraisal of pollution effects can be made. Toxicity tests at the level at which interspecific competition can occur and community repercussions be recorded are, however, vital. Few laboratories have attempted this type of work.[19,20] The Orielton group have already published their initial work[20] and most members of this Workshop will be familiar with their approach. Laboratory toxicity tests were an important part of the Orielton programme and were particularly used to choose appropriate toxic oils and dispersants for field trials. They were also important in obtaining data that aided interpretation of observed community response. No one at Orielton undervalued their importance in aiding the planning of full-scale field trials. Only with the field experiments, however, was it possible to make realistic ecological predictions from correlated laboratory-based information. An appropriate representation of the desired relationship between laboratory trials, field trials and true pollution investigations was given by Baker and Crapp:[4]

Toxicity bioassay on single species[2]
↓
Laboratory experiments on selected species and community[4] dominants
↓
Field experiments on communities and groups[10] of organisms
↓
Observation on situation in nature[21] following pollutions

The toxicity test at the top of this scheme is a long way from what is needed to interpret the situation at the bottom.[21] It is, of course, an invaluable part of the complete investigation and an excellent starting-point. There can be little doubt that much more effort is required at the community and eco-system level of response. Without this, ecological predictions based entirely upon laboratory work are likely to be invalid. Many examples exist in the literature forecasting a doom which never materialised.

I do not want to develop standard tests, for in my experience standard = static = no progress. What I hope we can achieve is some progress towards a code of practice for toxicity tests which will give guidelines to which all laboratories can turn. Such guidelines should cover everything from animal selection to the style of data handling, presentation and interpretation. I also hope that those conducting toxicity test programmes will recognise the limitations of the various approaches and methods.

In relation to toxicity-ranking tests, I suggest that owing to the wide differences in LD_{50}, etc., found between different laboratories using the same materials and test organism (often of two or more orders of magnitude), absolute figures are relatively meaningless. Such large sources of experimental error are associated with all methods that it would be more realistic to assign toxicities to a simple toxicity scale related to results obtained from a standard test material. Such a reference material should not be a commercial product but rather a readily available analytical quality hydrocarbon.[16]

I hope that this summary provides a foundation for an interesting and useful discussion that will continue after this Workshop itself has been concluded. All of us will have views on the many problems raised. Many biologists will disagree in varying degrees with the views that I have presented as a basis for our discussion. All of us, I know, will have justifications for our own lines of approach. If I have raised a storm of argument from which some sort of progressive re-evaluation emerges, then I shall have achieved my objective.

REFERENCES

1. Beynon, L. R., and Cowell, E. B. (eds.), *Ecological Aspects of Toxicity Testing of Oils and Dispersants*, Applied Science Publishers, London, 1974.
2. Wilson, K. W., 'Toxicity Testing for Ranking Oils and Oil Dispersants', this volume, pp. 11–22.

3. Crapp, G. B., 'Laboratory Experiments with Emulsifiers', in *The Ecological Effects of Oil Pollution on Littoral Communities* (ed. E. B. Cowell), Institute of Petroleum, London, 1971, pp. 129–49.
4. Baker, J. M., and Crapp, G. B., 'Toxicity Tests for Predicting the Ecological Effects of Oil and Emulsifier Pollution on Littoral Communities', this volume, pp. 23–40.
5. Nelson-Smith, A., 'Micro-respirometry and Emulsifier Toxicity', Field Studies Council, Oil Pollution Research Unit, Annual Report, 1969.
6. Bellan, G. L., 'Toxicity Testing at the Station Marine d'Endoume', this volume, pp. 63–7.
7. Swedmark, M., 'Toxicity Testing at Kristineberg Zoological Station', this volume, pp. 41–51.
8. Ottway, S. M., 'The Comparative Toxicities of Crude Oils', in *The Ecological Effects of Oil Pollution on Littoral Communities* (ed. E. B. Cowell), Institute of Petroleum, London, 1971.
9. Cowell, E. B., and Hainsworth, S., Observations on salinity tolerances in littoral gastropods (in preparation).
10. Crapp, G. B., 'The Ecological Effects of Standard Oil', in *The Ecological Effects of Oil Pollution on Littoral Communities* (ed. E. B. Cowell), Institute of Petroleum, London, 1971.
11. Baker, J. M., 'Seasonal Effects', in *The Ecological Effects of Oil Pollution on Littoral Communities* (ed. E. B. Cowell), Institute of Petroleum, London, 1971.
12. Wishart, J., and Sanders, H. G., *Principles and Practice of Field Experimentation*, Commonwealth Agricultural Bureau, London, 1955.
13. Fisher, R. A., On the mathematical foundations of theoretical statistics, *Phil. Trans. R. Soc.*, Ser. A, **222**, 309–68 (1922).
14. Ganning, B., and Billing, U., 'Effects on Community Metabolism of Oil and Chemically Dispersed Oil on Baltic Bladder Wrack, *Fucus vesiculosus*', this volume, pp. 53–61.
15. Spooner, M. F., and Corkett, C. J., 'A Method for Testing the Toxicity of Suspended Oil Droplets on Planktonic Copepods Used at Plymouth', this volume, pp. 69–74.
16. Wilson, K. W., Cowell, E. B., and Beynon, L. R., 'The Toxicity Testing of Oils and Dispersants: A European View', this volume, pp. 129–41.
17. Gunkel, W., 'Toxicity Testing at the Biologische Anstalt Helgoland, West Germany', this volume, pp. 75–85.
18. Shell Company Ltd, 'Quality Criteria for Aviation Fuels', Shell Report SPS/222, Aug. 1972.
19. Perkins, E. J., Some effects of detergents in the marine environment, *Chem. Ind.*, pp. 14–22 (1970).
20. Cowell, E. B. (ed.), *The Ecological Effects of Oil Pollution on Littoral Communities*, Institute of Petroleum, London, 1971, 250 pp.
21. Smith, J. E. (ed.), '*Torrey Canyon*' *Pollution and Marine Life*, Cambridge Univ. Press, 1968.

Discussion

Mr L. R. Beynon: A great deal of work has been done on some of the older dispersants like BP 1002. This material has been the joy of all biologists in the laboratory because its toxic effects on marine life are so dramatic. It is time, however, that test work should now concentrate on the modern low-toxicity dispersants and that less emphasis should be placed on the older toxic materials.

I am now going to call on Mr Andren of the Fisheries Resources Division of FAO in Rome to make a brief statement on some of the work that FAO committees have been doing and some of the recommendations which have been made.

Mr L. E. Andren (Fisheries Resources Division, FAO, Rome, Italy): I shall only summarise the conclusions drawn at a recent meeting that was held by an 'interdisciplinary expert group' which was formed to advise seven of the UN agencies on scientific problems related to marine pollution.

The major importance and function of this group is that it is intended to represent the views of scientists involved in studies on, for example, oceanography, chemistry, geochemistry, biology and fisheries, meteorology, etc. The subject of bioassays is not, of course, a typical agenda item for a group of this nature. It was felt in FAO, however, that when we were intending to establish working groups to deal in more detail with bioassay problems, it would be appropriate to have some guidance of a general nature from a group representing a very wide range of scientific interests. This guidance would prove useful for FAO policy development and for our directions to other specialist working groups that are being formed.

Of the working groups convened in 1973, one has a marine bias and consists of the group that is normally advising FAO on scientific matters in relation to fisheries (the Advisory Committee on Marine Resources Research). Another group advising FAO represents IABO (the International Association of Biological Organisations). Another group which will soon be convened will deal mainly with problems related to inland waters. We realise that there will be an overlap here with the European Inland Fisheries Advisory Committee (EIFAC) which has been dealing with such problems for some time.

The recommendations which I am going to cite now are again of a general nature as opposed to the many interesting, intriguing and detailed points that Mr Cowell raised.

The terms of reference of the GESAMP working group that met a month or so ago were:

(a) definition of the term 'bioassay';
(b) identification of the problems to be answered by bioassays;
(c) critical review of the tests available and their limitations;
(d) discussion of the utilisation of bioassays and other pollution parameters in the definition of water quality criteria and standards.

For the purposes of its deliberations, the working group adopted the following definition of the word 'bioassay'. It is *an experiment using aquatic organisms, or their individual organs, to examine the response (usually detrimental but not necessarily so) of substances or energy added to or subtracted from the water.* Also considered to be bioassays are *those experiments where organisms are fed or injected with substances.*

The report of the working group and the discussions of the plenary session concluded that although bioassays appear unlikely to be suitable in isolation for the establishment of water quality criteria to protect all uses of the marine environment, they will continue to provide vital information for such purposes. The group recommended that:

1. Attention should be given to the further development and use of bioassay procedures, particularly in the fields of pathological, biochemical and physiological effects.
2. Where it is necessary, for various reasons, to conduct acute toxicity tests, the continuation of the tests to evaluate threshold concentrations in addition to the TL values for various time periods should be encouraged.
3. Where possible, analysis should be carried out on the animals killed or surviving bioassay tests, particularly where bio-accumulation is suspected.
4. Further use should be made of experiments both in the laboratory and in the field to ensure the detection and assessment of synergistic effects of pollutants.
5. Encouragement should be given to an early conclusion to discussions with fisheries organisations such as EIFAC, preferably in collaboration with terrestrial and medical fields of interest, with a view to the production of an internationally agreed terminology for toxicological bioassay procedures.
6. Further attention should be given by GESAMP to the relative importance of bioassay results and other pollution parameters for the evaluation of water quality standards to protect different kinds of water use.
7. The continuation of bioassay tests to establish the dose/effect relationship for given marine organisms should be encouraged.
8. Investigations should be made to determine the extent of carry-over of pollutants during the course of desalination.
9. In relation to the incidence of biotoxin formation, the following studies should be encouraged:
   (a) the determination of the ecological factors which might be contributory causes to biotoxin formation;
   (b) standardisation of the methods of evaluating the presence of biotoxins;
   (c) forecasting the occurrence of biotoxin formation.

The report will be published with a related annex containing the detailed work of the working group.

Mr K. W. Wilson: I doubt whether the difference between 'doses' and LT_{50} values has ever been clearly defined in relation to the aquatic environment. The way we use these terms is that if you maintain animals in given concentrations until they die, that gives you the LT_{50}. If you expose several batches of animals to the same concentration and determine the effect time for killing them, this then becomes a 'dose'. For example, perhaps in one experiment you have a range of pollutant concentrations, and in each of them you measure the time taken for 50% of the animals to die. However, if in another experiment, as I explained yesterday, you have one pollutant concentration and examine the test animals frequently and plot percentage mortality against time, you can read off the 50% value. Since these animals have been in this concentration all the time, we term this an LT_{50}. If, on the other hand, you make up several batches of that same concentration and put animals in each of these, you can remove animals after different lengths of time and put them in clean sea water and see how they get on. Again, you can plot percentage mortality against the time that the animals were exposed to the test concentration, *not* the time that it takes them to die. You may find, in fact, that of the animals which were removed from this test concentration alive, 20% died two or three days later. From this sort of data you work out the 'effective dose' to kill 50% of the animals.

Mr A. Mann (Shell International Petroleum Co. Ltd, London, England): What units are these figures in? You cannot have a 'dose' because you are talking in terms of concentration.

Mr K. W. Wilson: We are defining 'dose' here as a concentration multiplied by time.

Mr A. Mann: But you do not know what quantity of material the animal has ingested, so that it is not strictly speaking a 'dose'.

Mr E. B. Cowell: It does not matter what term you use so long as every time you use it in public you define it.

Mr A. Mann: It does matter because LC_{50} has a specific meaning in other toxicity tests and there will inevitably be confusion if you use the same term for this type of test.

Mr K. W. Wilson: I agree, I am just pointing out how it is often used, I am not agreeing with the usage.

Mr L. R. Beynon: This discussion clearly indicates the need for the work being undertaken by GESAMP, which was referred to by Mr Andren.

Mr K. W. Wilson: Perhaps I may also be permitted to reply at this time to some of the observations made by Mr Cowell. First of all, let us consider the purpose of your toxicity test. Quite simply, the thinking behind the test at Burnham-on-Crouch stems from the *Torrey Canyon* incident and the experience with dispersant usage at that time. For example, we found that pilchard eggs were killed in some areas. In the laboratory, you may find that pilchard eggs are killed at 10 ppm of dispersant A in 4 days, whereas with another dispersant, B, pilchard eggs are not killed in 4 days at 1000 ppm. At least then you know that, given a choice, use of dispersant B is less likely to kill pilchard eggs, which is common sense. Therefore the test that we devised was designed to determine whether new formulations of dispersant represent an improvement on older dispersants. Here we used BP 1002 as our standard reference material, largely by accident rather than design. It is lucky perhaps that many other workers have also taken BP 1002 as a reference standard so that at least the results of work in different laboratories may be compared.

The level of precision of the test is important and, like Mr Cowell, we are really only concerned with differences in orders of magnitude. If a dispersant is only half as toxic as one of its predecessors, we do not count that as a significant improvement. We would be looking for a difference of two or three orders of magnitude to signify improvements.

It is very important to define the purpose of your test and, as already stated, at Burnham-on-Crouch we undertake tests for the purpose of ranking dispersants. For such tests, in my opinion, it does not really matter which species of animal you use, because each species tends to give the same order of ranking when related to our reference material, BP 1002. This means that the points raised by Mr Cowell regarding choice of representative species become redundant.

Regarding selection of test organisms with respect to size and age, in practical terms you may find that you have new dispersants to be tested every week. Therefore it is no use selecting an immature female brown shrimp when the water temperature outside is 12°C because it means that you can only do your tests on three days in the year. While these considerations are theoretically important, in practice we are forced into some sort of compromise. We find that variations in size, condition, etc., have very little overall effect on the LD_{50} value for ranking tests, remembering that we are looking for differences in orders of magnitude between dispersants.

Randomisation of treatment again falls into a similar category; it only affects the results slightly.

Mr E. B. Cowell: It all depends on the conditions in your laboratory. If you have a row of test tanks in a room which has a window at one end and a heater at the other, randomisation is extremely important. In fact you could not do an analysis of variance on the results with any degree of validity unless they had been randomised.

Mr K. W. Wilson: I agree. I had assumed that as many variables as possible were standardised, for example tank size, number of tanks, number of animals, test temperature throughout the year, etc. You do of course get different results if you undertake your tests at, say, 6°C or 20°C. At Burnham-on-Crouch the winter temperature may drop to 1° or 2°C. If you do a test at that time of year it does not give a very good comparison with tests undertaken in the middle of summer at 24°C. We use as a base line 15°C, and we look for effects on our test animals of taking them from their ambient conditions to this standardised temperature. Again, we find that differences are small. Other physical factors we also find to be of little significance, although we do not delude ourselves, for example, into thinking that we have a constant concentration of dispersant. We look upon this as an uncontrollable variable. Different types of dispersant behave differently under test, and it is impossible to standardise on one effective method of producing known concentrations of dispersants. One dispersant might be especially prone to evaporation, another to degradation, another to loss to the tank walls, and so on. Again, however, we find in practice that the importance of such effects is not large, remembering that we are looking only for very large differences in test results.

Mr L. R. Beynon: There appears to be some divergence of opinion between Mr Wilson and Mr Cowell, but also some agreement, since they both feel that ranking dispersants in groups is preferable to determining absolute values and putting too much emphasis on these. Do we have general agreement on this point?

Dr M. Swedmark: I certainly agree. I also feel that every laboratory undertaking toxicity tests must try to standardise on their techniques. We need a standard form for ranking, and LC_{50} value is both suitable and convenient. You must expect variations in LC_{50} values, but these are not so large as to discredit their use for ranking in groups. This method of ranking in groups is very valuable as a means of giving information to authorities who are faced with the practical task of dealing with oil pollution.

It is also possible, of course, to carry out more sophisticated tests. For example, we combine toxicity testing with effects on biological functions, and I do not agree with Mr Cowell that these two approaches clash. We keep our LC_{50} tests going tor 96 hours at a time, and as well as observing mortalities we note other effects. For example, observing the byssal activity of the mussel is a good means of differentiating between dispersants and enables one to predict how these animals will be affected by long exposure. Perhaps we should confine ourselves initially, however, to the practice of toxicity testing for ranking because this has more immediate practical value than the more complex tests involving biological predictions.

Mr E. B. Cowell: I do not really disagree with either Mr Wilson or Dr Swedmark, but I feel that we must state the limitations that can be placed on the methods and recognise what variables exist. It is far too easy otherwise for governments, local authorities, etc., to infer far too much from the results of toxicity experiments.

Mr J. Wardley Smith (International Tanker Owners' Pollution Federation, London, England): Part of the confusion which arises is because tests are carried out for a variety of purposes. A government wants to know whether a certain dispersant is highly toxic or has a low toxicity so that a decision may be made regarding its suitability for clean-up. The absolute value of LC_{50} means little in this respect. However, a manufacturer may be able to use absolute values of LC_{50} to determine trends in the toxicity of various formulations to assist him in developing new dispersants. The third type of test is a more academic one in which the various parameters have to be chosen very carefully to give the optimum answer on the type of information one is seeking.

Mr L. R. Beynon: The ranking tests are most useful to authorities because they tell them which materials are going to be least harmful in practice. Mr Cowell suggested five categories for ranking; I should like to leave out the fifth—'non-toxic'—because I believe that nothing is non-toxic if used in sufficient quantities. May I suggest for your consideration, therefore, that we might agree on the following rankings:

extremely toxic	LC_{50} up to 100 ppm
toxic	LC_{50} between 100 and 1000 ppm
moderately toxic	LC_{50} between 1000 and 10 000 ppm
slightly toxic	LC_{50} greater than 10 000 ppm.

Mr K. W. Wilson: I believe that the Warren Spring Laboratory has produced such a list. They designate an LC_{50} of greater than 10 000 ppm as 'practically non-toxic' and the other toxicity ratings descend in orders of magnitude, with the highest toxicity denoted by an LC_{50} of less than 100 ppm.

Mr E. B. Cowell: It is important in this context that one should have a reference material, to which one may return, which is always repeatable. We have been told that many people use BP 1002 for this purpose, but what would happen if it ceased to be manufactured? It would be far preferable to have an 'analytical reagent type standard'.

Mr L. R. Beynon: This is a very valid point. I sometimes feel that BP 1002 is only made nowadays for biological tests.

Mr A. Mann: Could I say particularly that if we have a standard material it has to have the same general chemical and physical characteristics as the substance under test. If we are looking at dispersants, the standard must be a surface-active material. There is no point in using a hydrocarbon as standard if we are really interested in surfactants, because we are likely to get a different mechanism of toxicity and this will invalidate the test. Again, it appears from Mr Wilson's remarks yesterday that the Burnham-on-Crouch test applies to conditions off-shore. In this context, if we are going to have a ranking we must say quite clearly that the ranking is not going to apply to beach-cleaning materials, for which a much shorter exposure time must be used.

Mr L. R. Beynon: We have come to a discussion of interpretation and application of test results. I should like to add another thought for consideration. Nowadays people tend to think of dispersants only in terms of their toxicity, but it is no use having a material which is 'practically non-toxic' if it is of no use as a dispersant.

Mr J. T. Daniels (Phillips Petroleum (Europe–Africa) Ltd, London, England): I think that Mr Mann introduced an important point when he mentioned the mechanism of toxicity, and there seems to have been a lack of research in this respect. Of prime importance to us, however, is the toxicity of the oil. If there is an absence of oil with some toxic properties there is no need to use dispersants, yet the majority of papers presented here have been concerned with the toxicity of oil/dispersant mixtures or with dispersants alone. What work has been done on the toxicity of crude oils on their own?

Mr L. R. Beynon: There has been a great deal of work done on crude oil toxicity, and perhaps I could ask one of the people from the Orielton Field Studies Centre to discuss this point with respect to the work carried out by Dr Ottway.

Dr J. M. Baker: We were supplied with about 20 fresh crude oils and tested them for toxicity ranking using the marine snail *Littorina littoralis* and saltmarsh plants. The same sort of toxicity ranking was obtained independent of the organism used for the test. In terms of what causes the toxicity, it was found that a high aromatic content of the oils correlated with high toxicity, and also that the physical nature of the oil is important. Light oils, being of a more penetrating nature, could get into the organisms and cause more damage than heavier, more viscous oils.

Dr M. Swedmark: The degree of emulsification of the oil is also very important. If you test the oil as we did, with the oil floating on the water or in big droplets, it does not appear to be as toxic as oil which is very well emulsified mechanically. In the latter case it was suggested that the 'toxic component' is better spread out in the water. We also found in our tests on dispersants and surfactants alone, which produce a decrease in surface tension and have an effect on gills, that a ranking order for animal groups for such materials was that fish were most sensitive, then bivalves, and that crustaceans were very resistant. But if we tested mechanically emulsified oil/dispersant mixtures, the ranking order changed: fish were not so sensitive, the sensitivity of bivalves increased, and crustaceans, especially, were very sensitive to emulsified oils. This effect must mean that there is another action of oil on the animals. In this case effects on the gills do not seem to be so important. Crustaceans have a waxy film on the carapace and it is possible that oils, which are lipophilic by nature, have a solvent effect on the wax film. This work suggests, therefore, that the choice of animals for toxicity testing is important.

Mr L. R. Beynon: I should like to return for a moment to Mr Daniels' question. One of the reasons why people have looked at the toxicity of dispersants alone is that five or six years ago the materials which were available were far more toxic than crude oils, so that by comparison the oil had little effect. This is no longer true. One must also remember, of course, that there is nothing that one can do about the toxicity of oil; if oil is spilt, one has to accept what has been spilt. On the other hand, one can do something about the toxicity of dispersants used for cleaning it up.

Mr J. T. Daniels: While that may be correct with respect to oil spills, there is another situation which may give us cause for concern. At Dubai and in the

Ekofisk field in the North Sea we have storage tanks where we are dissolving oil in water continuously. If we had a clearer idea of toxicity mechanism we could design storage tanks to make the area around our facilities more acceptable to marine life.

Mr L. R. Beynon: Commenting briefly on Dr Swedmark's remarks, choice of species is very important, not least because some are mobile and some are sessile, and some may be considered as somewhere in between. But the other big difference between the laboratory and the field test is that the sea is enormous, and one does not only get dispersion there but also dilution. This effect is very difficult to achieve in the laboratory.

Mr G. C. Silsby (British Petroleum Co Ltd, London, England): I think that the idea of classifying dispersants in a relatively small number of classes is good, but I do think that it is a bit early to start tying LC_{50} labels to them. We have only to refer to two of the papers here to see that the LC_{50} values awarded to two of the dispersants by different laboratories represented here are vastly different. Any attempt to assign ratings in terms of 0–100 ppm, 100–1000 ppm, etc., is pointless unless you have a standard test.

Mr L. R. Beynon: I agree. We need agreement between laboratories, but we also need agreement on whether LC_{50} determinations should be made over 24 hours, 48 hours, 96 hours or whatever.

Mr E. B. Cowell: The point of ranking in terms of four or five orders is that it really does not matter what differences a laboratory gets in LC_{50}'s in their own particular test. So long as they stick to the same test and the same base line they should produce the same rankings from very different LC_{50} types of determination.

Mr A. Mann: It is important not to give a set of figures for the rankings. Personally I have seen LC_{50} figures varying by a factor of 100 for different species on a 48-hour basis.

Dr G. B. Crapp: It seems to me that we have some agreement in ranking that it does not matter what species you are using. Does this not imply that the brown shrimp is an inefficient animal to use? We might get results a good deal more quickly, and we might cut down some biological variability, if we used what I shall call a 'universal biological thing', perhaps some sort of cell culture. Surely for ranking we cannot justify going out into the field and collecting animals, which is time-consuming, and possibly using a test which takes several days. We might be able to do it in a respirometer, perhaps on an algal culture, in a matter of hours.

Mr L. R. Beynon: This is an interesting point and relates back to what Mr Cowell was saying about Dr Nelson-Smith's work on yeasts.

Dr R. Johnston (Department of Agriculture and Fisheries for Scotland, Marine Laboratory, Aberdeen, Scotland): There is one very important point here. Marine algae have a reproduction time of perhaps $1\frac{1}{2}$ days, whereas for the cod that Dr Swedmark was talking about, for example, its reproduction time is one year. Therefore the cod which she is killing may be two or three years old.

Dr J. M. Baker: I think that the crucial thing to determine is whether it is true to assume that toxicity testing is independent of the species you use. I have seen evidence produced by Mr Wilson and at Orielton to suggest that this is so, but yesterday Mr Wilson also quoted American data produced by Dr Tarzwell which are contrary to this assumption. I am not entirely sure what this work was, and I wonder whether anybody else has any contrary evidence.

Mr A. Mann: Dr Tarzwell, in particular, has worked with brine shrimp. I feel sure, however, that it is dangerous to say that the choice of species is unimportant. Some species are very resistant.

Mr L. R. Beynon: Surely what we need for these ranking tests is a species which is sufficiently sensitive to differentiate between dispersants in terms of the broad bands of toxicity rating which we have been talking about. Several people have taken the view in other types of test that they should choose the most sensitive species possible, and have taken the Pavlovian attitude that any effect on the animal in terms of response is bad.

Mr E. B. Cowell: I still think that some of us are missing the point with regard to ranking tests. If you choose a sensitive species you will get a set of LC_{50} values which will give you a ranking order. If you choose a less sensitive species you will get a shift in the LC_{50} values but still get the same ranking order. Provided your test is standard, even if you use a yeast in a respirometer, it should be possible to allocate these same hard categories, although for very different criteria.

Mr L. R. Beynon: What I was suggesting was that you have to have a species which will distinguish between 'toxic' and 'slightly toxic'. It is no use choosing a species where every animal dies or one where no animal is affected no matter what dispersant is used.

Dr W. Gunkel: In ranking tests, would it not be good to go down to a very basic reaction and not to a reaction which comes from several other steps? Take, for example, assimilation of green algae. For a cell culture we have very basic reactions. Moreover this material would be available all the year round, which is also a very important point. We must remember, of course, that it would have nothing to do with ecology, but it could be useful for standardisation for ranking. It would do away with the problem of whether the dispersant had an effect on the gills or elsewhere. For standardisation it is important that in all cases the same reaction is influenced.

Mr L. R. Beynon: This is some support for Dr Crapp's point of view.

Dr G. B. Crapp: It is very easy for a biologist to get confused between the two objectives of ranking and ecological predictions. Biologists often feel that they must take a species which lives in an environment they know something about, with the vague idea that they can make some ecological prediction from the ranking test. If they are not careful, they may end up with a test which is fairly slow and inefficient for ranking and which still does not give enough information for ecological prediction. It is possible then to end up with a compromise test which is not really the best for either purpose.

Mr L. R. Beynon: I think we can agree on a statement that ranking tests on their own have nothing to do with ecological predictions.

Mr A. Mann: If ranking tests have nothing to do with ecology, of what value are they? If you are going to kill animals by some mechanism, is there any value in ranking a material by a method which does not take account of the mechanism which causes the effect?

Mr E. B. Cowell: It simply enables you to say that one material is likely to be better than another, but it does not give you any information about the kinds of interactions which may be expected in an environmental situation. An example Dr Baker gave yesterday is a very good one in this respect. You can put a lot of animals in a laboratory and kill the limpets, but this tells you nothing about ecology unless you know something about the limpet and its biology in the field. But if you want to know whether your new dispersant is better than the one which you made last year, then a ranking test is a very valuable tool.

Mr L. R. Beynon: I should like to return to something that Mr Cowell said earlier about the need for different types of tests, and Dr Baker's suggestion that one should have laboratory tests, field tests and examination of the natural situation (including pollutants). We have some other information as well. Fortunately or unfortunately, we do have oil-spill accidents in which clean-up action is taken, and we do, by and large, know the biological results of undertaking such clean-up, although perhaps not in sufficient detail. But we do have some actual field experience to say, 'The laboratory test showed this dispersant to be extremely toxic and this has been borne out; this test showed a dispersant to be only slightly toxic and this has been borne out.' Whilst such practical experience is important in the toxicity context, it is even more important on the question of efficiency of dispersants. If I remember rightly, the Warren Spring Laboratory devised a magnificently reproducible test for beach-cleaning efficiency—a very difficult task—and their enthusiasm for the test only waned a little when they discovered that the material which had been found to be most effective in the field had been rated the least effective in the laboratory.

Do we need agreement in this broad band of toxicity testing for ranking on a closer definition of LC_{50}? Should it be related to 24-, 48- or 96-hour time periods, or whatever?

Dr J. M. Baker: If you are just using an organism such as yeast for a ranking test, there is no need for LC_{50}'s or LD_{50}'s to be determined at all. You can do the test in an hour or so in a respirometer by measuring effects on respiration rates.

Mr J. T. Daniels: Surely that disagrees with the work done in Brussels which showed that you have different response times with different dispersants. The effect of a dispersant is very different after 24 hours from the effect after 48 hours.

Mr E. B. Cowell: This comes back once again to the confusion between ranking and ecological tests. The time does not come into a ranking situation at all.

Mr A. Mann: Surely it could.

Mr L. R. Beynon: You are suggesting that in 24-hour LC_{50} tests one could get one ranking in the extremely toxic, toxic, etc., bands, whereas 96-hour LC_{50} tests would give you a different ranking. Hopefully if you get into these very broad bands you might get away from that situation.

Dr G. B. Crapp: A point Mr Wilson made yesterday was that for any sort of efficiency in ranking experiments you should not use a test that only gives you an LC_{50} value. You should use an animal whereby you can note the death point and plot a graph from which you can read off the threshold and also the LC_{50}'s for 24, 48 and 96 hours.

Mr L. R. Beynon: Mr Wilson also seemed to be saying that if you go far enough in time then you will always get the same rankings.

Mr A. Mann: May I ask how you can use a respirometer test to compare a biodegradable with a non-biodegradable dispersant?

Dr J. M. Baker: You have to use an organism which is not capable of biodegrading the dispersant.

Mr A. Mann: How can you find whether or not this is true? How can you isolate this organism from the micro-organisms which will be present in the system anyway?

Dr W. Gunkel: There are two ways. Firstly, if you use green algae you are not likely to get degradation; these will synthesise using only CO_2 for the living factors. Secondly, you might use the penicillin technique in which you add a mixture of bacteria to the dispersant and after some time add a large amount of penicillin. All the active bacteria will be killed by the penicillin but not those which did not grow/degrade. After this you destroy the penicillin and extract even one or two bacteria out of perhaps 10 million and allow them to grow, thus isolating a strain which does not degrade dispersant.

Mr E. B. Cowell: We keep getting away from the purpose of ranking tests. If you consider the difference between biodegradable and non-biodegradable dispersants, the ranking test still only measures immediate toxicity; when discussing degradation products you may be measuring the toxicity of something else. There are many cases where biodegradation products are themselves toxic.

Mr L. R. Beynon: Do all the people concerned with ranking tests agree that they can discard their life's work and start again with basic organisms?

Mr A. Mann: This is only possible if the mechanism of toxicity does not involve some parameter which only occurs in higher animals. If your dispersant, for example, is a neurotoxin, then work on monocellular organisms will be of little use.

Mr K. W. Wilson: Once again, someone is trying to pretend that from a laboratory ranking test one is able to predict what is going to happen in the field. The ranking test only acts as a preliminary sorter of materials and is followed by a much more careful selection of test organisms as one moves into the field.

Mr L. R. Beynon: Basically, most of these laboratory tests, be they for ranking toxicity or efficiency, should at least 'separate the sheep from the goats'. I believe that you only get the really interesting results when, having discarded the 'non-starters', you really get down to the ecological situation with field tests. Ranking tests for toxicity and efficiency play an important role in initially screening the products, but they never replace practical trials.

Mr N. Robinson (Texaco Ltd, London, England): There seems to be a danger here of saying, on the one hand, that ranking tests are intended to give information to governments and other organisations and, on the other hand, that they are only intended to get rid of the non-starters. A disastrous situation could arise if a test indicated that a dispersant was practically non-toxic whereas, in fact, it was highly neurotoxic. The ranking order must be meaningful to the organisation which is making use of the information.

Mr A. Mann: If a ranking order is given to government, the limitations of the ranking order must be clearly stated.

Mr E. B. Cowell: It is up to ecologists to state the limitations which should be placed on interpretation of their results. If governments ask for other than a ranking test, then we have to think again regarding what test should be devised to meet their requirements.

Mr L. R. Beynon: Unless we have something which biologists feel is a meaningful ranking test, then the government will never have a meaningful ranking test. Surely on biological aspects governments must listen to the advice of biologists.

Mr J. Rogers (Department of Trade and Industry, Warren Spring Laboratory, Stevenage, England): Governments do take notice of biologists. The Warren Spring Laboratory has several dispersants per week coming in for assessment. The last test applied, assuming that a product meets all other requirements, is a biological one.

Mr L. R. Beynon: There have been isolated instances of governments who have formulated a policy and thereafter will only accept biological data which support that policy. By and large, however, governments sincerely want advice which will help them to formulate a reasonable and meaningful policy.

Mr E. B. Cowell: If anything useful comes out of this Workshop at all it should be an agreement that ranking tests are only the beginning of assessing the impact of a particular material on the environment. Biologists must make it clear that they have not done anything magical in carrying out ranking tests other than saying 'This is the order of things'. It is like sorting a pack of cards; it does not tell you how to play the game.

Mr L. R. Beynon: This is a fairly good analogy. At least if you can sort the cards into suits you are starting to get somewhere.

Dr W. Gunkel: If biologists are honest with themselves they must say that they do not fully understand 'natural ecology'. How then can we claim to understand ecology after we have added a toxicant to the system, thus creating 'unnatural ecology'. We cannot expect, therefore, that with our methods we can arrive at the end-point. We can only hope to be on the way to becoming a little more ecological.

Mr L. R. Beynon: Can we now consider tests from which ecological predictions may be made? Let us leave ranking tests.

Dr A. R. D. Stebbing: May I suggest one thing that one has to determine in a test from which ecological predictions are to be made? That is the lowest concentration at which a pollutant has a measurable deleterious effect upon organisms in the environment.

Mr L. R. Beynon: In this respect, are you defining 'deleterious' in terms of any change in the behaviour of the organism?

Dr A. R. D. Stebbing: Yes, for example in terms of depressed growth rate or depressed breeding.

Mr L. R. Beynon: Or in terms of increased growth rate or increasing breeding?

Dr A. R. D. Stebbing: Yes.

Mr A. Mann: That may not necessarily be deleterious.

Mr L. R. Beynon: I have met this argument elsewhere. Many people believe that any change is deleterious.

Mr A. D. McIntyre: Perhaps I could be allowed to say something of the approach which we make to our long-term studies of toxicity. We start with LC_{50} experiments on a series of particular organisms which we want to use, which are probably not group placed in a particular area. We then look at conditions in the field in an area in which we are particularly interested, perhaps in the vicinity of an effluent pipe or pipeline. This gives us two values: an upper value based on the LC_{50} limit, and a value of the actual concentration of a pollutant, which might be copper for example, which pertains in the field. We then design a series of long-term experiments, which may last for six months, using a series of concentrations between these two values. The experiments are not on particular organisms, they are on a food chain. We use large tanks, we have primary production going on, we have primary production utilised by prey which might be a bivalve, and this in turn is fed on by predators such as young fish. So, by looking at the effects on this simplified ecosystem over a period of six months or so of different concentrations of a pollutant, we can get some sort of idea of the effects of this pollutant on such things as mortality and growth rate on different parts of the food chain. One has to be very careful, however, in extrapolating from this situation to field conditions, but at least we know that within these levels something is happening and we can perhaps make suggestions regarding what might happen if the concentration of a pollutant doubled or trebled. Although we are chary of making these extrapolations, we feel that this is at least a useful approach.

Mr L. R. Beynon: There are several basic ways of looking at this sort of ecological problem either in the laboratory or in the field. These include looking at total populations, dominant species and sensitive species. These are three distinct possibilities, although there are also others, but has anyone any views regarding which attitude should be taken? Or should one take all three attitudes and examine them all?

Dr R. Johnston: If you are in a fisheries laboratory you are interested in the food chain, and you should undertake the sorts of experiments which Mr McIntyre has described.

Dr J. M. Baker: Ideally one has to turn to field experiments in the community in which you are particularly interested. First of all you have to decide which community or ecosystem you are concerned with—which system you want to make predictions about. We have divided the Milford Haven area into several systems for study—rocky shores, salt marshes, sand and mud flats, and so on. Field experiments are then best because you are concerned with the total population in an area, and you do not have to make any assumptions about what are the key species and what are the interactions. If you bring various animals and plants to the laboratory, even if you test them together, you do not have the whole population there. You have to make assumptions that they are the key species and you would miss interactions which are only observable in the field.

Mr L. R. Beynon: In the field, of course, you do have natural stresses imposed, for example by extreme weather conditions. In the laboratory it is possible, in theory at least, to eliminate such stresses.

Mr E. B. Cowell: Perhaps I could further illustrate Dr Baker's point. Some years ago I was working on a very strange plant community in some hill land in Scotland. We started by examining community dominance and ignored one or two species which were rather rare. Then, having removed some of the community dominants by using a herbicide, to our surprise some of the things which we had discarded as unimportant became dominant because they were able to fill the spaces which we had created. This illustrates that community interaction can never be fully understood by bringing only community dominants into the laboratory.

Mr A. Mann: This also illustrates that species selection is very important. Could we recommend that there should be three stages: the crude bioassay test for ranking, a simplified ecosystem in the laboratory, and field tests? The laboratory tests should be used for screening, so that eventually few materials need be field-tested.

Mr L. R. Beynon: Before going into this subject in more detail, perhaps we should think about why we are doing these tests and what ecological predictions we want to make. Let us consider the UK situation, for example. There are certain areas around the coast where one is not allowed to use dispersants because there is something special which we want to protect, perhaps a shell fishery. Is it any use in examining the effects of dispersants on these creatures in this type of ecosystem when the use of dispersants is banned? You cannot control where you accidentally spill oil, but you can control the use of dispersants. Again, on the shore, in many places one is told that clean-up may not be undertaken and that the oil must be left untreated to protect a local ecosystem, and the consequences of an oil-polluted shore must be accepted.

Mr E. B. Cowell: Returning to a point made by Mr Mann, this question of species selection seems very simple to a non-biologist. If you go to the Baltic, where Dr Ganning told us yesterday that there are 52 species, then conceivably you could do experiments on the lot within the course of two or three years. If you consider the British coastal ecosystems you are dealing with vast numbers of species. Ecologists have been saying for years, give us 50 years and enough research money and we shall begin to tell you the answers, but we do not have this sort of time scale available. We do not yet know which species are important in simple terms so we have to go to the field and look at community interactions. They will begin to indicate the important species which should be examined in the laboratory. The interactions are the subject of greatest importance. It may be a long time before we understand why they occur, but meanwhile we can at least observe what interactions do occur.

Mr L. R. Beynon: A very well-documented example of interaction occurred following the *Torrey Canyon* accident. Large numbers of limpets having been killed by dispersant cleaning, there was an invasion of *Entermorpha* spp. into the littoral zone.

Dr M. Swedmark: Even laboratory experiments can give you a lot of information about biological consequences if you study biological functions as we have done. You take simple functions such as shell closure, syphon retraction, and so on; they are protection or avoidance mechanisms. The animals which we study live in the littoral zone and they are adapted to very wide changes in environmental conditions involving temperature, salinity, etc. Such animals have adapted to protect themselves against extreme conditions and we have seen in our aquaria that these animals react similarly to pollution. Oysters and mussels, for example, will close their shells tightly when there is sufficient concentration of a pollutant to cause reaction, whereas animals such as clams and cockles will burrow more deeply into the mud or sand. These are the main reasons why relatively little damage has been caused in the littoral zone in many cases following oil spills. Littoral zone animals can protect themselves for short periods. Animals living in the sub-littoral zone and below do not have as much need for protection mechanisms because the environment is more stable, and they are perhaps more sensitive to pollution. Sinking oil as a method of pollution abatement is therefore dangerous because the animals living on the bottom are so sensitive and unprotected.

The next step in the behaviour pattern, which is reduced activity, is indicative of a much more dangerous situation. Here the differences in species resistance are important as distinct from avoidance reactions. In a simple laboratory ecosystem we had prawns, shore crabs and hermit crabs, the prawns being most sensitive and the shore crabs most resistant. At sublethal concentrations of dispersant the activity of the prawns decreased, they did not react to food, they lost their defence reaction, they did not swim away from the crabs, and they were eaten. Such sublethal concentration studies in aquaria can at least enable one to anticipate that similar situations will occur in the field. One has only to study simple biological functions and there is no need to become involved in complex physiological processes.

Mr L. R. Beynon: There are two situations coming in here: gross insult and low-level pollution. In the oil-spill situation and the use of dispersants, assuming that cleaning is carried out correctly, what you have in the littoral zone is a high concentration of dispersant for a short period. The chronic oil pollution situation is very different. Here you may have very low concentrations for a very long period, so that one is looking for a different type of effect.

Dr M. Swedmark: We have carried out long-term effects of low levels (in the range 0·1–1·0 ppm) of household detergents. These concentrations are well below the short-term lethal threshold, but kills did occur after many months exposure. It was also interesting that when we allowed the laboratory temperature to vary with the natural ambient temperature, we had the greatest mortalities in May–June when the temperature had increased from the low winter levels. This indicates that temperature stresses can make animals more sensitive to low levels of pollution. This illustrates the need for carrying out similar long-term experiments with very low concentrations of oil alone.

Dr G. B. Crapp: Dr Swedmark's laboratory experiments only indicate what *might* happen under natural conditions using sublethal levels. An an example, let us consider the dogwhelk *Nucella lapillus*, which is a predator on barnacles and mussels. Perkins at Strathclyde University has shown that, after exposure to a sublethal dose of dispersant, the feeding and growth rates are depressed. What is not known is how much this matters in the field. Bryan at Plymouth studied dogwhelks after the *Torrey Canyon* incident and suggested that there was some evidence of a decrease in growth rate after pollution, but this was no more than could be accounted for by immediate narcosis induced by the dispersants used. There was no reason to suppose that decreased growth rate had been observed on the shore. Even when an animal is affected in this way there are other ecological factors to be considered. Let us suppose that a dogwhelk survives pollution and let us suppose that its feeding and growth rates are suppressed. What we found on a typical barnacle- and limpet-dominated shore was that limpets died, algae invaded the shore, and under the subsequent thick algal cover the barnacles on which the dogwhelk feeds died. Within about a year or two there was no food left for the dogwhelks and it did not seem to matter whether their feeding rate had been depressed or not.

Mr K. W. Wilson: This is a very important point. In the way that we have defined exactly what we are trying to do in our ranking test, it is important that the people who are concerned with sublethal and ecological testing define what they are trying to do. It seems to me that they are trying to find out what happens when dispersants are sprayed in the field, and the simplest approach would be to go and spray in the field and see what happens.

Dr G. B. Crapp: You need all these laboratory investigations into sublethal effects, delayed mortalities, etc., to help explain your results in the field. We are lucky that when we do get oil spills we can in a sense experiment with dispersants in the field during oil clean-up.

Mr E. B. Cowell: Community interaction is a very important additional factor when you are working in the field. For instance, you may have an animal which will recover from quite heavy dispersant treatment, and from a laboratory test you may deduce that it is in no danger. But if that animal gets dislodged from a rock because of dispersant treatment, it may be exposed to predation from a quite unexpected source.

Mr L. R. Beynon: A similar unexpected result could also occur, of course, if dispersant treatment results in a kill of an animal's food source.

Mr E. B. Cowell: Yes. If you have a resistant predator but its prey is sensitive, you might deduce from a laboratory test that the predator was unaffected. But in the field if its food no longer existed it would die just as surely as if you had killed it with dispersant.

Dr W. Gunkel: I am very much in favour of field experiments, but you have to remember that the type of experiment described by Dr Baker applies only to the benthic and littoral communities. The most productive areas are not in the littoral zone; all the primary production occurs in the sea. The water is subject, of course, to movement as a result of tides, currents, etc., and it is very difficult to follow a particular water body in field experiments. This is a limitation of field experiments which we have tried to overcome in two ways. Firstly, we use an 'underwater laboratory' with divers who live in the environment and see what is going on there. Secondly, as I described yesterday, we have an ecological laboratory where we have constructed plankton towers which are 10 metres high and have a diameter of 5–8 metres. In this way we are able to study a large water body; this could prove to be a way of closing the gap between laboratory and field experiments.

Mr L. R. Beynon: Dr Gunkel has raised the point that a great deal of primary production occurs offshore. Perhaps I could ask him whether he agrees with what appears to be the generally accepted Western European thesis that it is better to deal with oil at sea rather than to wait for it to come ashore.

Dr W. Gunkel: The main problem in the North Sea is that, driven by wind and current, oil will find its way ashore within two or three days. Any dispersant action must therefore be taken very quickly. Many people agree, however, that chronic oil pollution of the sea is a bigger threat than a tanker disaster. The former contributes some millions of tons per annum to oil pollution, whereas the largest tanker disaster yet encountered released only 100 000 tons of oil to the sea. But if it is impossible to pick up oil slicks from the water, it may be best to let them come ashore and deal with them mechanically there.

Mr. L. R. Beynon: There was a classic case of an accident in the North Sea in 1966 to the tanker *Ann Mildred Brovig*. A lot of oil was released and, while people were setting up emergency schemes to deal with it, most of it just disappeared, so other factors were coming into play. With reference to Dr Gunkel's suggestion of letting the oil come ashore, it is a pity that we do not have an ornithologist here today. Had we had one present, I am sure that he would have been objecting strenuously to the suggestion that no action should be taken against floating oil but that it should be allowed to drift ashore. Such action, or rather lack of action, would maximise the damage to sea birds. The third point I should like to make is with respect to mechanical cleaning of shores. I do not know what damage would be inflicted on marine life by huge bulldozers collecting sand and dumping it. Neither do I understand the philosophy of people who ban the use of low-toxicity dispersants for cleaning rocks for fear of harming marine life, but suggest sand-blasting or steam-cleaning them instead.

Mr H. J. Marcinowski: With respect to the *Ann Mildred Brovig* accident, it has been calculated that it would have been cheaper to allow the oil to come ashore before tackling it than to take action to deal with it at sea. One cannot generalise, however, on this subject because a great deal depends on the type of coastline threatened. Again, the decision to clean may depend on whether a beach is used by tourists and whether an accident occurs during the tourist season.

Mr L. R. Beynon: It is not merely a question of cost, of course. If you did deal with all the oil at sea, so that none of it came ashore, it would put a lot of biologists out of work.

Dr W. Gunkel: Returning to the *Ann Mildred Brovig* accident, very little of the oil was treated at sea and yet coastal pollution was limited to one small area in Denmark. The oil must have been degraded or dispersed naturally in the body of the sea. In such cases, therefore, is it not best to do nothing?

Mr L. R. Beynon: There is an accepted view, certainly in the UK, that if you spill oil in the open sea and continuous observation shows that it is unlikely to come ashore, that it is unlikely to pollute important fishing grounds, and that it is unlikely to invade important bird-breeding areas, then it is permissible just to continue to observe it. There are risks involved, however.

Dr M. Swedmark: Most of the oil which may be spilled in the North Sea is likely to find its way to the Swedish coast because of the prevailing winds and currents. There certainly was an increase in pollution of our coasts during 1972.

Dr W. Gunkel: But there were no tanker accidents in the North Sea during 1972. This illustrates that it is probably the relatively low level of chronic pollution that is more important than the disaster situation.

Mr L. R. Beynon: We seem to be wandering a little from the point of discussion. Should we consider how major accidents and chronic pollution fit into our concepts of toxicity testing to make ecological predictions?

Mr J. Wardley Smith: I agree with Dr Baker that you should do experiments in the actual environment, but what worries me is what you should take as your indication that harm has occurred. Some people would agree with me that the effect on Cornwall of the *Torrey Canyon* disaster was negligible. Even six months after the accident many people were predicting this, and today there is no traceable effect. In this context, considering experiments on small plots of ground, what do you take as your limiting condition? This is the sort of question to which the government needs an answer so that, faced with the task of beach clean-up, they can be reasonably sure that they will not cause irreparable damage.

Mr E. B. Cowell: As an ecologist, I should like to point out that there was enormous damage to Cornwall after the *Torrey Canyon*, but that it has recovered. Overall it was a minor perturbation in a major ecosystem.

Mr J. Wardley Smith: Yes, so that the object of the clean-up exercise, which was protection of the tourist industry, was successful, and the environment has not suffered permanently.

Dr G. B. Crapp: This brings us back to the criteria which Dr Stebbing described for environmental deterioration. In a way, we are not justified in looking for any absolute standard of pollution damage, because all pollution damage is relative. It depends who you are, what you do, and what you are prepared to put up with.

Dr A. R. D. Stebbing: Was enough known about the flora and fauna of Cornwall to be sure what the effect of the *Torrey Canyon* was? It is only now, through working parties on long-term monitoring, that people are beginning to appreciate how little we know of the biology of our coastlines. Perhaps one of the best-documented pieces of our coastline is that around the Marine Biological Association laboratory at Plymouth. In this area, certain changes in population are becoming apparent. I believe that the numbers of species and their abundance are decreasing without our becoming aware of it because we have not been collecting the right data over the years.

Mr L. R. Beynon: I think that we would all agree that there is an unfortunate lack of background data on what one might call the 'normal' ecological situation around our coasts.

Dr R. Johnston: Government scientists have to advise the government about the probable effects of pollution, and the mercury problem, for example, is not nearly so reversible as the oil pollution problem. Trying to find the wisdom to assess what the effect of a new pollutant might be is extremely difficult. One can carry out both short- and long-term tests, but how is it possible to preclude disasters like the methyl mercury tragedy or the thalidomide tragedy in the medical field? We have to take chances, but let us ensure that we make the risks as small as possible.

Mr E. B. Cowell: This raises another question with respect to toxicity testing. It was stated at the FAO Conference in Rome in December 1969 that it is the persistence of a toxicant which is of prime importance rather than its immediate toxicity.

Mr L. R. Beynon: We have talked a lot about the need for doing field trials and tests and undertaking studies of ecological systems, but we have also agreed that there is a need for laboratory tests. Some of the variables listed by Mr Cowell were chemical aspects, dispersion (maintenance of concentration), selection of materials for standardisation, and selection of test organisms. Are we in a position now to consider the laboratory methods in detail?

Dr E. N. Dodd: I should like to make two points with respect to the need for standardisation. Yesterday our attention was drawn to synergistic effects between oils and dispersants, both in Mr Wilson's and in Dr Swedmark's papers. By and large, the presence of dispersants results in a smaller oil droplet size than would result from dispersion of oil alone. Can biologists tell us whether oil droplet size is a significant factor in this type of testing? If it is, then we should attempt to standardise on droplet size. One approach might be to standardise on the mixing energy. The other point I should like to raise concerns the question of whether a standard hydrocarbon should be used in place of crude oil. Do we really know what are the toxic fractions in crude? We know that aromatics are toxic, but if the refining process used in preparing the standard material removes the toxic fractions, does not any toxicity test using the standard become a waste of time?

Mr L. R. Beynon: Two questions have been raised, one regarding the standardisation of emulsification techniques and one regarding standardisation of test oils.

Dr A. R. D. Stebbing: I should like to consider the medium in which one carries out tests. The constituents of natural sea water are not consistent with time; for example, pesticide run-off has an annual cycle. One may therefore be performing experiments in which there may not only be an annual cycle but also an increasing or decreasing concentration of a substance with time. This is why, in our laboratory, we standardise on artificial sea water.

Mr L. R. Beynon: Sea water may also vary, of course, from location to location. Salinity variations are a good example of this. May we discuss now the possible need for standardisation of sea water in addition to the earlier questions of emulsification and test oil standardisation?

Mr E. B. Cowell: My point with respect to a 'standard oil' is that it does not even have to be an oil. You need a standard material to which you can relate the toxicity of other materials. In the context of dispersants, for example, you should be able to relate the effects of BP 1002 with those of a standard material which is always obtainable.

Mr L. R. Beynon: A reference standard for comparison purposes is acceptable, but we also have the difficulty, which is associated with ecological testing or with relating laboratory testing to ecological knowledge, that we have to use materials which are important in practice. We may also need, therefore, a 'standard polluting oil'. There is a lot of controversy, for example, regarding whether one should use a 'fresh' crude oil or a 'weathered' crude oil when undertaking tests both in the laboratory or in the field. It is certainly true that the properties of crude oils change very rapidly when they are spilled on the sea.

Dr R. Johnston: Standard sea water may be useful in testing oils but it has little value when undertaking tests on, for example, trace metals.

Mr L. R. Beynon: Difficulties related to standard sea water also occurred following the *Torrey Canyon* disaster when laboratories were studying the formation of 'chocolate mousse'. It was extremely difficult to produce such water-in-oil emulsions using synthetic sea water, but very easy using natural sea water.

Mr A. Mann: Why was this so?

Dr R. Johnston: It is almost certain that artificial sea water made up from analar chemicals will contain much more trace metals than natural sea water; there are also many other reasons because sea water contains so many traces of so many different substances and organisms.

Dr W. Gunkel: In microbiological experiments it is common to obtain 'stable' sea water by allowing natural sea water to 'age' for perhaps three months.

Mr L. R. Beynon: Nobody has expressed a view on the need for a standardised emulsification technique.

Dr R. A. A. Blackman: I agree that it is important to standardise on the particle size of an emulsion if you are testing an animal in a range of different oils and dispersants, but the important particle size is relevant to the animal which you are testing. One should not attempt to standardise over a range of animals.

Mr L. R. Beynon: It should also be relevant to what happens in practice. One can get a complete range of droplet sizes dependent on the efficiency of dispersant usage. Perhaps one should assume that the people who are going to use dispersants will use them properly. It is not realistic to ban their use because they may be used improperly. As a crude analogy, one does not ban the motor-car from the road because some drivers make mistakes and accidents result.

Mr J. J. P. Dick (British Petroleum Co. Ltd, London, England): With regard to emulsification, Dr Dodd seemed to imply that it was relatively easy to obtain a standard emulsion—which implies a standard droplet size—by controlling the mixing energy supplied. This is an over-simplification. So far as practical measurements on ships are concerned, the difficulty in producing standard droplet sizes is the main problem in developing a precise method for measuring oil concentrations in discharge water. Droplet size is very important with respect to monitors utilising fluorescent techniques. If the biologists solved the problem of producing emulsions with standard droplet sizes, it would be of great value to instrument manufacturers.

I should like to make another point with respect to absorption of oil by tubing and other parts of experimental apparatus. In tanker measurements on gas concentrations in cargo tanks it is also very important to use tubing which does not preferentially absorb different components in the gas. We have largely solved this problem by using nylon tubing.

Mr L. R. Beynon: Taking your points in reverse order, nylon tubing is certainly very good in this respect, and it is also inert so that it does not tend to lose material to your experimental system. Going to the other extreme, rubber tubing is absolutely useless when working with oils and dispersants.

With respect to emulsification, it would be nice if we were able simply to put in enough mixing energy to get uniform, say $1.5\,\mu$, oil droplets, but this is not easy to do in practice.

Dr R. Johnston: Many workers in the biological field have been able to obtain topped crude from the petroleum industry. Would it be possible to obtain from the oil companies a stable emulsion of known particle size for use in toxicity experiments?

Mr L. R. Beynon: No! One may be able to get fairly reproducible emulsions if one standardised the emulsification techniques and always used the same materials under exactly the same conditions. It is extremely difficult, if not impossible, to obtain the right sort of emulsion which is stable for a very long time.

Dr M. Swedmark: The method used in our laboratory for producing emulsions was very simple and seemed to be satisfactory. We use a propeller mixer and get quite uniform, small droplets.

Dr B. Ganning: This may be true when using one oil and one dispersant at one concentration, but as soon as you change your materials you will get different droplet sizes resulting from your applied mixing energy.

Mr L. R. Beynon: I have seen propeller-type mixers used in toxicity tests to produce sufficient agitation to keep oil in suspension. I am not sure, however, what the effect of the resultant turbulence was on the marine creatures under examination.

Dr M. Swedmark: In our tests the mixing is applied so that the animals are not affected.

Mr K. W. Wilson: Certainly we do not have any problems with our control animals. We have a 20-litre container with a power-driven stirrer. The animals appear to be all right and feed well over a four-day period. I should not like to go further than that.

Mr L. R. Beynon: If you produce uniform emulsions which are very cloudy, how do you know that the animals are alive?

Mr K. W. Wilson: You stop the motor periodically to check the animals.

Mr E. B. Cowell: At the risk of sticking my neck out again, I should just like to comment that as a biologist who has recently come into contact with a great number of people from the oil industry, who are chemists, I find that they are very critical of almost all toxicity test reports. Chemists do not believe that the emulsions are anything like stable enough to justify the conclusions which are drawn from the tests.

Dr M. Swedmark: On the other hand, if you use a flow method and produce the emulsion continuously, there is not a problem.

Mr E. B. Cowell: My experience is that chemists do not agree on this point.

Dr M. Swedmark: When we did the tests we photographed the emulsions to see that we had about the same size of droplets. I do not see any difficulty.

Mr L. R. Beynon: The difficulty is lessened, as pointed out earlier, if you always use the same oil and one dispersant. As soon as you start switching oils and dispersants, you have all sorts of other factors coming in, because oil is a very complex substance.

Dr M. Swedmark: We used marine diesel oil, heavy fuel oil and Oman crude oil and they functioned perfectly.

Mr L. R. Beynon: One source of marine diesel oil presumably? You see, diesel oil can vary enormously, within specifications, depending on the type of refinery, type of crude and various other factors.

Dr M. Swedmark: I know that there can be a lot of variation, but if it functions with Corexit, which is considered to be an indifferent dispersant, why should it not work with others?

Mr L. R. Beynon: As I said, it will vary according to the oil. A good example of this is the water-in-oil emulsion which some crudes make very easily; Kuwait is a classic example. Other crudes do not produce these emulsions.

Is there a feeling that we ought to use a fresh or a weathered crude in a test? If a weathered crude is preferred, then how weathered? A material which we have supplied on several occasions is topped Kuwait, for example, which I believe we 'topped' to 250°C; this is something like the effect of one to two days at sea.

Mr J. J. P. Dick: It seems to me that your work would benefit from more correlation of results. Someone should take a toxic chemical, such as benzene, and relate its effects on marine organisms to, for example, Kuwait crude. Is anyone thinking of doing this type of work?

Mr L. R. Beynon: Thank you for reminding me of what used to be a hobby-horse of mine. The big point which very few people know about all these tests is their reproducibility. What happens if another laboratory, using the same equipment and test materials, does the same tests? How reproducible is the test? This is a point I was making in the USA several years ago. People were proposing very complicated and expensive tests with many oils, dispersants and organisms. No one was thinking of testing the tests. From a scientific point of view this is a very important point. We did get agreement that the American authorities would conduct a co-operative test programme. When I checked on their progress some eighteen months afterwards—and it must be three years since the initial discussions—they had not arrived at a satisfactory test. As far as I know this is still the position.

There is little or nothing published on the reproducibility of toxicity tests.

Mr A. D. McIntyre: This has certainly been done in the case of heavy metals. There have been quite a number of materials circulated to various laboratories which have checked the toxicity of metals such as copper and mercury. The results corresponded with each other and have been published. Obviously this is the kind of thing which you are thinking of for oils.

Mr L. R. Beynon: We have been dubious on the efficiency side. Your hosts here today are the members of the IP Oil Dispersants Working Group. This Group set out to develop standard methods for both efficiency and toxicity testing. The efficiency one we thought was very easy. People had been doing these tests for about ten years and making all sorts of predictions from them. As soon as we went into a co-operative test programme, and we kept it up for four years, we were unable to achieve a reproducible meaningful test.

Mr K. W. Wilson: This is what I have had to do recently with regard to dispersants. We are doing a 'ring test' where different laboratories are using their own particular tests. They are given a series of dispersants and have to produce a ranking order.

Mr L. R. Beynon: I think I would simplify it even further by ensuring that everybody used the same tests. I should like to see a written test method, in the way that IP or ASTM methods are written, where all materials, equipment and conditions are specified. Then ask, say, six laboratories to do them and see if you get the same answers. Of course you will not get identical results, but let us at least get an estimate before we start applying results, in great detail, to the field situation.

Dr R. Johnston: I thought we had agreed that we were not going to try to extrapolate from short-term bioassays to the field situation.

Mr L. R. Beynon: I was looking at the part which says 'tests from which ecological predictions can be made'.

Dr R. Johnston: In that case it is no use testing something which is abundant in the tropics to assess its effect in the Arctic.

Mr L. R. Beynon: Agreed.

Dr R. Johnston: I think that I would agree with Mr Cowell that people should be allowed freedom to choose the type of test to suit the ecology of different areas.

Mr L. R. Beynon: I would agree again. What I should like to know is how reproducible any particular, appropriate test actually is.

Dr R. Johnston: I think one should specify the problem, but it is a matter of pressures. If you can find sufficient time you can do random tests, but it takes a lot of effort, especially in long-term ecological studies. I feel that if anyone is studying yeasts in respirometers, it is reasonable to undertake a large number of tests, but I do not see how one can realistically expect such testing in ecological studies.

Mr L. R. Beynon: The point which I was making—at the risk of being insulting to some of those present—is that biologists seem to be more prone than some of the other scientific disciplines to assume that the results they obtain have absolute meaning.

Mr K. W. Wilson: We have repeated tests on particular animals with one dispersant and always found agreement within 10–20%. This is our repeatability.

Mr L. R. Beynon: Similarly with efficiency testing, Warren Spring found that an operator who was used to doing the tests could obtain repeatable results. Nevertheless, as soon as we tried to repeat the tests using the same equipment in another laboratory, it was hopeless.

Mr E. B. Cowell: I want to come back to something which worries me and concerns biologists as a group. This is their arrogance about their work; in particular, their attitude to this question of emulsions. We have our expertise as biologists, which I hope the chemists will recognise, but it is quite wrong of us to ignore the warnings of chemists, who are specialists in the chemistry of emulsions. It is dangerous to assume that a standard type of test produces a standard type of emulsion, in terms of stability and droplet size. I think that an experimental programme is needed on the effect of droplet size, emulsions breaking in equipment, and so on. One example is a material which is used for tank cleaning, which contains two types of surface-active agent, one for making emulsions and the other for breaking them. Testing would never give a stable emulsion, even in continuous-flow equipment, for the moment the mixing energy decreases, the emulsion will start breaking. I believe that we must take heed of chemists who regularly say, 'Sorry, but we can't accept your biological data, because you do not know what you are saying on the subject of emulsions.'

Mr L. R. Beynon: I would support the need for closer contact between the disciplines. There used to be a tendency, particularly among biologists, not to want to talk to anybody who worked in industry. I am happy to say that this attitude is changing, although not as quickly in North America as in Europe.

Another odd thing is that biologists, government people, universities and so on always seem surprised that the oil industry knows anything about oil. I always remember the case of a very eminent man, who shall remain nameless, who came to me and asked for a sample of Kuwait crude which he said he wanted to analyse. Not only could we write a book on the subject, but we have actually done so and are willing to make this information available.

Mr N. Robertson (Texaco Ltd, London, England): It may be completely irrelevant, but on the subject of emulsions there is an IP standard test for emulsifiable cutting oils. It may not be applicable here, but it has been standardised and found to be reasonably good over a certain range.

Mr L. R. Beynon: If the meeting thought that it was worthwhile, I could go back to the Institute of Petroleum and recommend that they set up a working group to try to find an acceptable standard method for forming dispersions. We could use the work that Mr Robinson just mentioned as a starting-point.

Dr R. Johnston: I do not think that it is true to say that biologists ignore the advice of chemists. It is an old-fashioned attitude that biologists work on their own. I believe that nowadays most biologists are working in multi-disciplinary groups and we are very conscious of advice we get from people working in different fields. If the chemists are not satisfied with some aspects of the tests, we should be delighted to receive suggestions for improvements.

Mr L. R. Beynon: I think that Mr Cowell's point was that the chemists should be given a *chance* to comment.

Dr R. Johnston: I gathered from something Mr Cowell said that there was some reason for investigating the effect of different droplet size on ecological systems. I do not think that your working party ought to be answering this type of problem, which is quite different from the question of emulsion preparations and stability.

Mr L. R. Beynon: I think that both are interrelated because if you can prepare several standard emulsions you can investigate the effects of droplet size. If you could, for example, prepare emulsions with droplet sizes of 1, 20, 50 and 2000 μ, it would allow investigation of their effects on marine creatures.

Mr N. Robinson: It must be remembered that droplet size increases as emulsion is left standing. Temperature also plays quite a big part; for example, boiling normally breaks an emulsion and freezing can also break certain types of emulsion.

Mr A. Mann: Are we not putting too much emphasis on the need for standardisation? I think we know that different dispersants produce different droplet sizes. Should we not specify the mixing energy which we must use and accept the toxicity results we get with the resulting different droplet sizes?

Mr L. R. Beynon: Specify an energy rather than a droplet size?

Mr A. Mann: Yes.

Dr R. Johnston: Would a simple solution be to show a microscope picture of the emulsion so that people can see what the droplet sizes were?

Mr L. R. Beynon: You can certainly do this. It is, however, not easy to determine droplet size distribution in liquids since they tend to change. For example, you have to ensure that your microscope light source does not heat the emulsion.

Mr E. B. Cowell: I know nothing about the subject, but is it not possible to use Millipore filters, or other kinds of filters, to regulate the particle sizes entering your equipment?

Mr L. R. Beynon: No. If you use something like Millipore filters, the prime way they work is by adsorption. Furthermore, the idea that a 5μ filter has 5μ holes is a common fallacy. In practice, filters have a considerable range of pore sizes which makes their behaviour very complex, even with solids.

Dr W. Gunkel: There are now filters available which are made from plastics in which the holes are 'shot' using a laser or some other electronic means. Consequently these filters have a very narrow range of pore size. They have been on the market for several months, and sizes of 5, 6 or $6\cdot5\mu$ are available.

Mr L. R. Beynon: In the case of oil droplets, filtration is complicated by the tendency of the droplets to stick to the surface. Filtration is not simply dependent on the hole size. Furthermore, you can change the size of an oil droplet by filtration, or distort it to enable it to pass through a hole.

Mr N. Robinson: It has been shown statistically that a nominal 5μ filter of the type which has just been mentioned has an absolute cut-off of around 75μ. Admittedly this is for solids, not oil droplets. The shape of particles is also a factor, since some are long and thin while others are short and fat.

Mr L. R. Beynon: I spent three to four years of my life looking at this problem. Even with spherical particles, which we sized absolutely, with 99% within $0\cdot5\mu$, it is a very complex problem.

Dr W. Gunkel: Could this not be done using the Coulter counter technique?

Mr L. R. Beynon: A great deal of work has been done using the Coulter counter, particularly in assessing the efficiency of filters. It works very well for uniform-size particles but there are problems when you have a wide range of particle sizes. We found in such cases that the Coulter counter never adequately replaced the tedious, visual microscope count.

Mr K. W. Wilson: A danger with doing toxicity tests is that you assume you will get similar response curves for all dispersants on your test animals. If you are getting different responses from your test animals to different dispersants, a comparison of these dispersants becomes invalid anyway. The usual observation is that when you add a dispersant to an oil, the general pattern of toxicities does not change. Therefore it does not really matter if you are comparing the dispersants alone, or the dispersants with oil. You are simply looking at toxicities in a slightly different, and very often contracted, range.

Dr B. Ganning: Can anyone say how large a difference in droplet size is needed to give a significant difference in toxicity? We have now been discussing for some time the problems associated with producing a stable emulsion with a well-defined droplet size. How important is this?

Mr E. B. Cowell: I think the answer is that we do not know, and we should.

Dr R. A. A. Blackman: It depends on the animal. If you are talking about a filter-feeder, then the size of droplets which you provide will determine whether the oil is ingested or rejected.

Mr L. R. Beynon: One small point. If we ever get round to a standard dispersant for comparison purposes, let us not take one of the old, very toxic ones.

Mrs C. van der Wielen: Would it be possible to produce standard oils by fractionating into a large number of different cuts?

Mr L. R. Beynon: It is quite difficult to arrange to obtain one type of oil as a standard and to ensure that it is stored so that no changes occur. It is not practical to do the same thing for a large number of different cuts from each oil.

Mr C. Carpenter (Admiralty Oil Laboratory, Cobham, England): I think the idea is that you should initially split your 45-gallon drum into, say, 1-litre containers so that you are not repeatedly opening the large container.

Mr A. D. McIntyre: I believe we are saying that you should always have a constant liquid/vapour volume ratio. Is this right?

Mr L. R. Beynon: Furthermore, you should preferably sample without opening the container.

Mr J. J. P. Dick: There seems to be a feeling of pessimism creeping in, because of all the difficulties with this work. This is surely true of any discipline. People should continue as they are going, but should put more effort into correlating the effects of crude oil and specific fractions within the crude. The information will then be available to devise a satisfactory test method. I do not, however, see any laboratories volunteering to take up this correlation work, and perhaps this is where the working group should be active.

Mr L. R. Beynon: I am not sure what sort of a working group this would be. Certainly not an IP one, as we do not have the correct expertise.

Mr J. J. P. Dick: Such a group might be best in a university where the different disciplines are within easy reach of each other. I think, however, that it requires an effort on somebody's part to initiate such work—if it is felt to be really necessary.

Mr L. R. Beynon: Ladies and gentlemen, we have come to the official time for closing our Workshop. It only remains for me to thank you all for attending and for contributing so readily, and in such an interesting manner, to our discussions over the last two days. It is clear that a great deal of work remains to be done before all the complex problems related to the toxicity testing of oils and dispersants may be claimed to have been solved. This will involve a multi-disciplinary approach between all those who are interested in preserving our environment, whether they work in government, industry or independent research laboratories. For my part, acting as chairman of a biological Workshop, with such distinguished participants, has been a fascinating and very worthwhile experience.

11

The Toxicity Testing of Oils and Dispersants: A European View

K. W. WILSON

(Fisheries Laboratory, Ministry of Agriculture, Fisheries and Food, Fisheries Laboratory, Burnham-on-Crouch, Essex, England)

E. B. COWELL

(British Petroleum Co. Ltd, England)

and

L. R. BEYNON

(Institute of Petroleum, Dispersants Working Party, England)

THE PURPOSE OF TOXICITY TESTING[1]

The ultimate aim of toxicity testing is to predict the effects of a toxic substance on natural communities of animals and plants. Tests are used to explore two areas of variability, that which is found in the relative toxicities and modes of action of different substances, and that which is found in the complexity of natural biological communities and their environment. A single investigation is not suitable for studying both these areas; thus, a test which enables us to rank different compounds in order of their toxicity, quickly and accurately, is not usually very useful for making ecological predictions, whilst ecological studies are generally unsuitable for making comparisons of relative toxicities. This is reflected in the recognition of an interface between the acquisition of laboratory data and the prediction of ecological effects.[2] This is an extremely important point and cannot be overstated. Much of the confusion that surrounds toxicity tests arises from the misapplication of results, and all too often ecological damage is predicted on the basis of results from experiments which are manifestly not designed to provide such data.

A scheme of scientific investigation relating the two areas of variability can be represented:

(*a*) Simple bioassay with a single species

(*b*) Detailed bioassay with several species (dominant, most abundant, etc.)

(*c*) Laboratory investigation of sublethal effects

129

(d) Field
  experiments
  on a
  community

(e) Effect on
  a natural
  community

Detailed investigations at any one level cannot both describe and explain the effect of oil and oil dispersants on a community, and without this information it is impossible to predict the potential risk of a similar pollutant on the same community. Different sources of pollution and different communities require different laboratory approaches; thus, predictions of effects from results of long-term experiments at low concentrations of oil or dispersant are almost meaningless with respect to localised oil spillage. Similarly, acute toxicity tests are unlikely to give any indication of the long-term effects accruing from a continuous low-level discharge. Even long-term tests do not give information on long-term effects that accrue from interspecific competition and community interaction.

Initial bioassays are relatively easy to conduct and provide a means of ranking oils and dispersants on their toxicity under standardised test conditions. Equally, the description of the effects of accidental spillages is limited only to the timing of such occurrences. Field experiments, and more particularly laboratory studies on behaviour, etc., require careful interpretation.

TOXICITY TESTING FOR RANKING OILS AND DISPERSANTS

The aim of any ranking system is to express the result of a toxicity test by a single measurement and to use this to rank the toxicity of a group of substances. These initial ranking tests are useful when deciding which material to use in a field test for, owing to the large size of field experiments, it is clearly not feasible to repeat them with many different oils. More importantly they enable large numbers of dispersants to be screened, and those which represent a significant advance in terms of reduced toxicity to be identified. It is this type of information that is required by governments, oil companies, dispersant manufacturers and others to assess the relative values or risks involved in the use of various materials against other materials which have already been used in practice. These latter materials, of course, provide convenient reference points.

Unfortunately, considerable confusion has arisen because different workers have produced toxicity figures for the same material which differ by two or even three orders of magnitude. It is not possible to assess how much of this variability is due to differences in technique, test species, temperature, etc., since results for a common reference material are usually lacking. Quoting results of toxicity tests in absolute concentrations thus becomes meaningless. It was generally agreed at the Workshop on Toxicity Testing of Oils and Dispersants[1] that it is more meaningful to allocate such toxicity results to five categories of effect related to values obtained with reference materials.[3] These references should preferably be universally available, analytical reagent grade, pure hydrocarbons.

The importance of a degree of standardisation in the conduct of tests and in the presentation of results has been reiterated many times, but despite the recommendations of Doudoroff *et al.*[4] and APHA[5] among others, Sprague[6] concluded that the wide variety of approaches that still persisted had led to results which were often very difficult to compare. He pointed out that it was not uncommon to find lethal levels of one toxicant varying by a thousandfold. and he urged that it was 'time to organise the reasons for such variation'. Differences of use in terminology and in the collection and treatment of experimental data have added unnecessarily to this variability. An acceptable code of practice is now an urgent requirement.

Collection and Statistical Treatment of Data

In his review on the measurement of pollutant toxicity to fish, Sprague[6] reported that, broadly speaking, there are two procedures of acute toxicity testing in current use. In the first, mortalities in several concentrations of the substance under test are recorded only at fixed time intervals, usually 24, 48 or 96 hours, and the concentration lethal to half the animals after these times is interpolated from these data. In the second approach, the time to death of each individual animal is recorded and the time taken to obtain 50% mortality is calculated for each concentration. This latter procedure is recommended since it provides considerably more information from an equal number of animals. The use of standard statistical techniques to describe experimental data is valid only if the distributions of survival times, or the transformed survival times, are normal. One method of examining the normality of the transformed data is to examine them mathematically for goodness of fit. A more straightforward technique is to plot the data graphically, expressing percentage mortality as probits; the transformed survival times then lie on a straight line if they are normally distributed. The suitability of a logarithmic time/probit transformation for time mortality responses has been shown on numerous occasions since the detailed descriptions of Bliss.[7] Litchfield[8] describes how the basic statistics of the line can be obtained directly by plotting the original data on log-probability paper.

It is possible to describe fully the response of the population in terms of mean and variance. Considerably more accuracy is gained by measuring the average rather than a minimum or maximum response since the extremes of the distribution are so dependent on the size of the tank population.

Deviations from the straight-line relationship are encountered frequently. Apparent changes of slope at the extreme ends of the distributions are especially common and are doubtless symptomatic of the small numbers of test animals used. Truncated curves are usually explained in biological terms, viz. that the concentration is insufficient to kill the more resistant individuals of the distribution. In dispersant testing such changes in slope are often indicative of losses of toxin occurring during the experiment through, for example, evaporation, or of increases in the toxicity of a solution with time as may occur if surfactants are degraded to more toxic components.

In an experiment involving several different concentrations, estimates of the median survival time are made for each of these concentrations, and by plotting these times against their respective concentrations it is possible to establish a toxicity response curve. The suitability of logarithmic scales for both time and concentration axes has been commented on by numerous workers, although the semilogarithmic and reciprocal transformations have also been used.[9] It is extremely important to establish the toxicity response curves since fundamental differences between the shapes derived from two dispersants can indicate a difference in their mode of toxic action. Such differences have considerable bearing on the evaluation of the ecological risks arising from the use of dispersants. In extreme cases, attempts to rank dissimilar dispersant formulations are invalid.

From the toxicity response curve it is possible to derive one single measurement to describe the toxicity of a dispersant under particular test conditions, usually the concentration that is lethal (LC_{50}) to 50% of the test organisms after a given period. Most commonly used is the 24-hour LC_{50}, 48-hour LC_{50} or 96-hour LC_{50}. These fixed times are entirely arbitrary and, as they markedly influence the value of the LC_{50}, Sprague[6] recommends, as the most meaningful parameter with which to specify the effect of a toxin, the median lethal threshold concentration. At this concentration the toxicity ceases to be affected by further exposure.

Comparison of Dispersant Toxicities

If the toxicity response curves of several dispersants tested under the same conditions are compared, it is apparent that their relative toxicities are not constant with time and it follows that rank orders established at different times differ significantly. The standard procedure adopted at the Ministry of Agriculture Fisheries Laboratory, Burnham-on-Crouch, is to compare median lethal concentrations of dispersants at 48 hours (48-hour LC_{50}) and the median lethal threshold concentrations; the 48-hour LC_{50} is retained to allow comparisons with earlier determinations made at this Laboratory.[10,11] It is important to note that after 48 hours little change occurs in the rank order of most dispersants. Furthermore, reference to the effects of abiotic and biotic variables indicates that much of their influence is exerted on the lower portions of the toxicity response curve, with the threshold varying only slightly.

At present there are insufficient data for enough dispersants to allow comparisons of techniques. However, it is also implicit that any one standard technique should produce the same ranking order of a number of dispersants irrespective of the test species. The toxicity for ten dispersants given by Portmann and Connor[10] can be rearranged to produce a rank order with respect to their standard test species, the brown shrimp (*Crangon crangon*) (Table I). This order can be compared with those given by the other species using the rank correlation coefficient, r, where:

$$r = 1 - \frac{6 \sum d^2}{n(n^2 - 1)}$$

n = number of ranks
d = difference between any two rankings.

The rank order of all these species of crustacea showed close agreement, but significant differences existed between these and the bivalve mollusc, *Cardium edule*. However, the toxicities of these ten dispersants ranged over only one

TABLE I

A Comparison of the Rank Orders of Ten Dispersants to Four Species of Marine Animals[10]

Dispersant	Crangon 48-hr LC_{50}	Rank order			
		Crangon	Pandalus	Carcinus	Cardium
Slickgone 2	(3·5)	1	1	4	4
BP 1002	(5·8)	2	3	1	8
Slickgone 1	(6·6)	3	2	6	5
Gamlen OSR	(8·8)	4	7	3	2
Essolvene	(9·6)	5	5	2	7
Polyclens	(15·7)	6	4	5	9
Cleanasol	(44·0)	7	8	7	3
Slix	(119·5)	8	6	8	1
Atlas 1909	(120·0)	9	9	9	6
Dermol	(156·0)	10	10	10	10
r		—	0·879	0·818	0·152

order of magnitude and they should be viewed in the wider context of available dispersants. If the rank orders of *Crangon* and *Cardium* for dispersants which cover this wide range of toxicities are compared, there is very good agreement (Table II).

Baker[12] came to a similar conclusion when using saltmarsh turves in an unheated greenhouse to test the relative toxicites of oils.

Recovery from oil treatments was assessed by harvesting the turves two to five months after oil treatment, and obtaining a dry weight for healthy vegetation.

TABLE II

A Comparison of the Rank Orders of Ten Dispersants for *Crangon* and *Cardium*[9]

Dispersant	48-hr LC_{50}	Crangon Rank	Cardium Rank
Slickgone 2	(3·3–10)	1	3
BP 1002	(3·3–10)	2	5
Gamlen OSR	(3·3–10)	3	1
Cleanasol	(33–100)	4	2
Atlas 1901	(100–330)	5	4
Dermol	(100–300)	6	6
Polycomplex A	(100–330)	7	7
BP 1100	(1000–3300)	8	8
Corexit 7664	(3300–10 000)	9	9
BP 1100X	(> 10 000)	10	10
	$r = 0·843$		

Subsequent assessment of this type of test showed that it was very laborious, time- and space-consuming for ranking oils, and yielded relatively little additional information of use in making ecological assessments.[13] The data obtained, for example, on relative toxicities of crude oils did not differ significantly from those published by Ottway,[14] who ranked crude oils by an easier test using *Littorina littoralis*. In fact, as a general observation, relative toxicity information seems independent of the test organism involved, and the criteria for choosing experimental organisms should therefore be convenience and speed.[9,13] For example, Nelson-Smith[15] used yeast respiration rates in assessing dispersants.

TOXICITY TESTING FOR ECOLOGICAL PREDICTIONS

Two basic procedures have been described: laboratory studies and those which incorporate field studies.

Laboratory studies cover a wide range of approaches, from the determination of lethal levels in the short term to the study of effects at sublethal levels with a complete (and often uncritical) rejection of the determination of lethal levels.

Bellan et al.[16] determined short- and long-term lethal effects and conducted long-term tests to obtain quantitative data on the effects of detergents (dispersants?) on various stages of the life cycle of the polychaete worm, *Capitella capitata*, in small containers and under static conditions without oxygenation. Swedmark et al.[17] and Swedmark[18] adopted a similar technique but using a continuous-flow apparatus in an attempt to maintain the concentrations of oil and dispersant at a nominal level throughout the experiment. Lethal concentrations (96-hour LC_{50}) were derived for various selected marine species and, to determine the sublethal and chronic effects, a wide variety of biological functions were studied. The lowest or threshold concentration of toxin which affected different behaviour patterns was calculated. The authors concluded that from these 'ecological threshold concentrations' important conclusions on the ecological consequences of oil pollution could be made. Thus, the biological significance of the increased activity resulting from short exposure to oil dispersants was interpreted as an avoidance response and that this reaction guaranteed the survival of the individuals in situations of short exposure. They considered that the impairment of biological functions which followed with continued exposure to the dispersant after this period of increased activity was disadvantageous, and in the natural habitat, where the competition for food, space, etc., is important, the most resistant species is given an advantage, for example where the prey is selectively affected compared with its predator.

Recently Ganning[19,20] has demonstrated laboratory techniques which, he claims, would indicate the loss of production in a plant and an animal community exposed to sublethal pollution under field conditions. The daily oxygen requirement or production of microscopic algae was related to the level of oil pollution in his experimental tanks; a change of net daily oxygen production under normal conditions to one of a net requirement in a polluted situation being correlated to loss of primary production. In the brackish water amphipod, *Gammarus oceanicus*, the male and female enter into a praecopula prior to ovulation by the female. Ganning has demonstrated that this association is very prone to disruption by sublethal levels of oil and oil dispersants, and he concludes that, after such 'divorces', fertilised eggs cannot be produced and a decrease of the population ensues.

Attempts to establish the response of single species in a complex environment were made by Perkins,[21] who dosed littoral species in the laboratory before returning the marked animals to their natural habitat on the rocky shore. Using this technique, he showed that the periwinkle, *Littorina saxatilis*, when treated with a toxic oil dispersant (BP 1002) at doses of less than one three-thousandth of the 24-hour LC_{50}, died at a greater rate than controls for up to 22 weeks after the treatment.

Similar results were obtained for another winkle, *Littorina littorea*, and for the dogwhelk, *Nucella lapillus*. Furthermore, it was suggested that BP 1002 inhibited growth in *L. saxatilis* and *L. littorea* at less than one three-thousandth and one four-hundredth of their respective 24-hour LC_{50}.

This approach of relating laboratory and field experiments has been extended to cover the effects of oil and oil dispersants on rocky shore communities[22,23] and saltmarsh plant communities.[24,25]

In his investigations, Crapp first designed a simple laboratory toxicity test to determine the relative susceptibilities of different species to the same dispersant, BP 1002. Simultaneously, information was collected in the field after pollution incidents. It was not feasible to design a field experiment which was on a large enough scale to be little affected by migrations and small enough to cover a fairly homogeneous and easily counted community, and most of the field mortalities were assessed on accidentally polluted shores. In those cases a large area of shore was affected, but accurate pre-pollution data were not usually available and the intensity of dispersant application was difficult to determine. Some small-scale field experiments were made to obtain more accurate mortality figures.

BP 1002 was used in most experiments, for it was a widely used product at that time, and it was fairly similar in toxicity to most other products.[26,27,28] Animals were exposed to various concentrations of BP 1002 in sea water for one hour, after which they were thoroughly rinsed in sea water and left in this to recover. All these species retracted into the shell on encountering even very dilute solutions of the dispersant, and it was necessary to keep them for a five-day recovery period in order to determine which had recovered and which had died.

Crapp[23] found it convenient to describe the susceptibility of each species in terms of five categories, namely:

1. Very resistant: the one-hour LC_{50} is greater than 5×10^5 ppm. Examples: *Monodonta lineata* and *Littorina littorea*.
2. Resistant: the one-hour LC_{50} lies between 5×10^4 and 5×10^5 ppm. Examples: *Littorina saxatilis rudis*, *Nucella lapillus*, *Gibbula umbilicalis* and *Chthamalus stellatus*.
3. Moderately resistant: the one-hour LC_{50} lies between 5×10^3 and 5×10^4 ppm. Examples: *Littorina littoralis*, *Balanus balanoides* and *Elminius modestus*.

4. Moderately susceptible: the one-hour LC_{50} lies between 5×10^2 and 5×10^3 ppm. Example: *Mytilus edulis*.
5. Susceptible: the one-hour LC_{50} lies between 5×10 and 5×10^2 ppm. Example: *Patella vulgata*.

The critical part of this study was the relating of laboratory results to observations made of the effects of pollution and clean-up in the field, and this proved to be the most difficult part to carry out satisfactorily. Field observations were based on both accidentally and experimentally polluted shores.[23] A very important point was that animals may not be killed by dispersants, but affected in other ways. Littoral snails all retracted into their shells on encountering dispersant, and were dislodged from the rock surface, whilst other species remained attached. Thus, the numbers of these snails found after the spillage depended not only on how many survived but also upon how many were able to return to the shore after a period of being washed and rolled about. This initial disappearance, followed by the return of at least part of the population, has been observed after several cleaning incidents.[23,26,29] Of course, factors other than dispersant cleaning also affect mortalities.

Enough information was collected, however, on polluted shores to show that the relative susceptibilities determined in the laboratory were reflected in the field mortalities.

The experience gained from shore cleaning made it possible to differentiate between various situations involving the use of BP 1002.

1. When the dispersant was used sparingly and was washed away with large quantities of water, even the 'susceptible' limpet *Patella vulgata* was not severely affected. Populations of other species were sometimes depleted, but the grazers remained sufficiently numerous and active to prevent invasion of the shore by green and brown algae (see below). Recovery took place fairly rapidly and ecological relationships on the shore were not greatly disturbed.
2. Increased use of BP 1002 resulted in very heavy mortalities in *Patella vulgata*. 'Moderately susceptible' species suffered some mortality, and littoral snails retracted into their shells and were dislodged from the shore. Some of these molluscs survived and returned to the shore. Green and brown algae invaded the shore, but this development was restricted by the grazing activities of any surviving limpets, together with other herbivorous molluscs.
3. Heavy use of the dispersant for several successive days resulted in very heavy mortalities among the 'susceptible', 'moderately susceptible' and 'moderately resistant' species. In the more extreme cases heavy mortalities also occurred among the 'resistant' species. Many polluted shores cleaned to this extent were subsequently invaded by green and brown algae.[26]

This approach can give some idea of what the results of laboratory toxicity tests mean in ecological terms, and it is possible to use laboratory tests of new dispersants to make predictions on the ecological effects of their use.

Baker treated eight experimental plots in each of three different saltmarsh communities with Kuwait crude oil and was able to group the plant species in order of their relative tolerances.

Group 1 (very susceptible.)
Shallow rooting, usually annual, plants with no underground storage organs quickly killed by a single oil spillage.

Group 2 (susceptible)
Shrubby perennials with exposed branch ends which are badly damaged by oil.

Group 3 (susceptible)
Filamentous green algae. Though filaments are quickly killed, populations can recover rapidly by growth and vegetative reproduction of any unharmed fragments or spores.

Group 4 (intermediate)
Perennials which usually recover from a spillage or up to four light experimental oilings, but decline rapidly if further oiled.

Group 5 (resistant)
Perennials which have a competitive advantage in vegetation recovering from oil, due to fast growth rate and mat-forming habit.

Group 6 (resistant)
Perennials, usually of rosette habit, with underground storage organs (*e.g.* tap roots). Most of them die down in winter.

Group 7 (very resistant)
Perennials of group 6 type which have in addition a resistance to oil at the cellular level and have survived 12 successive monthly oilings.

Baker[13] gives an example of the post-pollution changes that occur in dominance as a result of the different recovery rates of oiled plants. The number of oil spillages that could be tolerated before the vegetation was all killed was also important. In a grazed zone, dominated by *Puccinellia maritima*, more than four monthly spillages resulted in persistent bare mud. Seasonal effects of oiling were also clearly demonstrated. Thus, winter oiling substantially reduced germination of seeds, and this was particularly noticeable in the case of annuals. Spring oiling reduced the flowering of many species, and thus reduced seed production later in the year. As with Crapp's study,[30] the results of the field experiments were compared with observations following oil spillages whenever possible.[31] The sets of data were found to correspond well.

RELATIONSHIP BETWEEN LABORATORY AND FIELD EXPERIMENTS

Simple bioassays provide comparative rankings of oils and dispersants but on their own they do not give any information that is useful in ecological terms. It is important that this limitation is accepted. Field studies, on the other hand, especially when considered in conjunction with laboratory experiments, have proved useful for such predictions. The role of the laboratory experiments alone in assessing long-term lethality or sublethal effects of oils and dispersants must be considered to be of indeterminate status. There are considerable methodological and technical problems associated with these tests as well as the overriding difficulties of interpretation or correlation with field situations.

Many workers embarking on such studies have selected 'representatives' of various types of organism. There are some reasons to doubt the validity of selecting a particular species to represent a taxonomic group, for there is a voluminous literature which shows adequately that no one species can adequately represent a group of species. The same argument applies to trophic level representatives and to animals chosen on the basis of feeding type. At best these representatives can give information only on the mechanisms of the reaction and types of behavioural responses within each group. They do not indicate the level at which other representatives would respond. Even within one species the animals selected for the experiments may not be representative of the population. In many cases it is possible to consider such variables as developmental state, sex, size and age (the two are not synonymous), but the population will be heterogeneous for many other important characters. No doubt selection will occur at capture, but all other variability should be reduced by randomly assigning the animals to the experimental treatments. There are numerous other variables which the investigator must recognise; for example, the period of acclimatisation in the laboratory, the test temperature, salinity, pH, oxygen tension, the number of animals per tank, food and level of feeding.[3]

Serious doubts have also been expressed on the probability of success of keeping oil concentrations and particle size distributions constant in experimental chambers. Static tank methods with or without agitation inevitably suffer from losses of the toxic material through evaporation and degradation, and continuous-flow methods designed to overcome these problems have their own intrinsic drawbacks. Many solvent-based dispersants attack plastic and rubber tubing, releasing toxins into the test tanks.[3]

Even if these problems are overcome, one can do no more than speculate on how effects determined in the laboratory relate to field situations. Care must be taken to observe whether an animal's behaviour is such that factors other than the oil or dispersant toxicity will affect its survival in the field. An obvious example of this is the behaviour of the littoral snails studied by Crapp.[30] These retract into the shell on encountering even low concentrations of dispersant.

The parameters of inactivation can be defined in the laboratory. Perkins[21] has determined the thresholds of inactivation in several snails, and Crapp[22] found that the rate of recovery was characteristic for each species and could be statistically defined in terms of its relationship with mortality. However, other factors affect field mortalities. The retracted snail may roll about or be swept away from the shore, and the numbers that recover will depend, among other things, upon the intensity and duration of dispersant use. The animals will be unable to avoid predators whilst inactive, and the abundance and activity of predators such as birds may be only loosely linked with the intensity of pollution. The animals may be carried by water movements into habitats where they cannot recover or return to the shore, and this will also be unrelated to pollution. This kind of behaviour can obviously upset ecological predictions. In the studies described by Crapp it was the return of some species to the shore that was significant rather than their original disappearance.

Animals which survive dispersant treatment may suffer an impaired vitality reflected in reduced rates of feeding and growth. Again this has been shown in experimental work; for instance, Perkins[21] found depressed growth rates in *Littorina saxatilis*, *L. littorea* and *Nucella lapillus*, and Crapp[22] found a reduced rate of feeding in *Nucella*, after exposure to BP 1002. However, little is known about how serious such sublethal changes are in natural communities. Bryan[29] studied the growth of *Nucella* after the *Torrey Canyon* spillage and found that the reduction in growth was no more than would be expected because of the period of inactivation.

Much more serious were the changes in community structure that were the result of observed mortalities. The activities of each species affect the biology of others, and if a proportion of the shore population is removed, then the conditions affecting settlement, growth and mortality change. The most striking example of this was the appearance of an algal canopy on shores where the herbivorous limpets had been killed, and many species, notably the barnacles, were greatly reduced in abundance under the weeds. Limpet spat began to settle on the rocks. As their numbers increased, fewer algal sporelings escaped grazing, and the appearance of the shore returned to normal as the older weeds died.

Particular attention has been given to shores dominated by limpets,[23] for most of those studied were dependent on limpet grazing for preventing algal invasion. In these cases it was found that there was little difference between the ecological consequences of the 'increased' and 'heavy' categories of cleaning with BP 1002 which have been described earlier. In both cases an algal canopy developed, and under this many species which survived 'increased' category cleaning were greatly reduced in abundance; thus, after two or three years the two kinds of shores appeared to be very similar and recovered their normal community in a similar way. This is an excellent illustration of the general principle that, if a key species is particularly susceptible, then even a moderate level of pollution or toxic dispersant application may have the most drastic effects.

The spores and larvae that settled and grew on polluted shores, once cleaning was over, were probably almost entirely derived from other, unpolluted shores, and the effects of dispersant were doubtless minimised by the fact that pollution was not continuous and treatment affected only a part of the littoral ecosystem. The affected shore was largely recolonised by the planktonic recruitment from plants and animals of more distant areas. The nature of the recolonisation depends primarily on the suitability of the shore for settlement and survival rather than on how many young the survivors can produce. Species with no planktonic dispersal phase in the life cycle will form an exception to this rule, and will be particularly vulnerable in terms of their ability to recover from pollution incidents.

CONCLUSIONS

1. It is necessary to design toxicity investigations to answer specific questions.
2. There are adequate tests available to produce comparative rankings of toxic materials in standard form. The basic principles which should be observed in carrying out these tests in the aquatic environment have been laid down repeatedly.
3. The choice of organism for ranking tests is unimportant. It should be selected for laboratory convenience, simplicity and speed of test.
4. In ranking tests, results should be related to standard reference materials.
5. Long-term laboratory tests suffer from disadvantages in methodology and technology. This is particularly true of tests involving oils and dispersants because of their chemical and physical properties.
6. Laboratory tests cannot be used in isolation to predict the ecological outcome of pollution incidents, although they are of assistance in interpreting effects observed in the field.
7. It is important that laboratory studies on several species are correlated with field studies at the community level.
8. Field experiments conducted on one type of community do not permit predictions to be made of ecological effects on dissimilar communities.
9. Results of work at community level are sadly lacking.
10. The design of all experiments, whether conducted in the laboratory or field, and the presentation of results obtained, should conform to accepted statistical practices.
11. There is an urgent need for the development of internationally acceptable laboratory codes of practice for toxicity testing oils and dispersants.

REFERENCES

1. Beynon, L. R., and Cowell, E. B. (eds.), *Ecological Aspects of Toxicity Testing of Oils and Dispersants*, Applied Science Publishers, London, 1974.
2. Edwards, R. W., Future research needs, *Proc. Roy. Soc. Lond.*, Ser. B, **177**, 463–8 (1971).
3. Cowell, E. B., 'A Critical Examination of Present Practice', this volume, pp. 97–104.
4. Doudoroff, P., *et al.*, Bioassay methods for the evaluation of acute toxicity of industrial wastes to fish, *Sewage and Wastes*, **23**, 1380–97 (1951).
5. APHA *et al.*, *Standard Methods for the Examination of Water and Waste Water including Bottom Sediments and Sludges*, Am. Pub. Health Assoc., New York, 12th edn., 1965.
6. Sprague, J. B., Measurement of pollutant toxicity to fish. I: Bioassay methods for acute toxicity, *Water Research*, **3**, 793–821 (1969).
7. Bliss, C. I., The calculation of the time–mortality curve, *Ann. Appl. Biol.*, **24**, 815–52 (1937).
8. Litchfield, J. T., A method for rapid graphic solution of time–percent effect curves, *J. Pharmac. Exp. Ther.*, **97**, 399–408 (1949).
9. Wilson, K. W., 'Toxicity Testing for Ranking Oils and Oil Dispersants', this volume, pp. 11–22.
10. Portmann, J. E., and Connor, P. M., The toxicity of several oil-spill removers to some species of fish and shellfish, *Mar. Biol.*, **1** (4), 322–9 (1968).

11. Portmann, J. E., 'The Toxicity of 120 Substances to Marine Organisms', Shellfish Information Leaflet No. 19, Fisheries Laboratory, Burnham-on-Crouch, UK, 1970.
12. Baker, J. M., 'Comparative Toxicities of Oils, Oil Fractions and Emulsifiers', in *The Ecological Effects of Oil Pollution on Littoral Communities* (ed. E. B. Cowell), Institute of Petroleum, London, 1971.
13. Baker, J. M., and Crapp, G. B., 'Toxicity Tests for Predicting the Ecological Effects of Oil and Emulsifier Pollution on Littoral Communities', this volume, pp. 23–40.
14. Ottway, S. M., 'The Comparative Toxicities of Crude Oils', in *The Ecological Effects of Oil Pollution on Littoral Communities* (ed. E. B. Cowell), Institute of Petroleum, London, 1971.
15. Nelson-Smith, A., 'Micro-respirometry and Emulsifier Toxicity', Field Studies Council, Oil Pollution Research Unit, Ann. Rep., 1969.
16. Bellan, G. L., 'Toxicity Testing at the Station Marine d'Endoume', this volume, pp. 63–7.
17. Swedmark, M., Braaten, B., Emanuelsson, E., and Granmo, Å., Biological effects of surface active agents on marine animals, *Mar. Biol.*, **9**, 183–201 (1971).
18. Swedmark, M., 'Toxicity Testing at Kristineberg Zoological Station', this volume, pp. 41–51.
19. Ganning, B., and Billing, U., 'Effects on Community Metabolism of Oil and Chemically Dispersed Oil on Baltic Bladder Wrack, *Fucus vesiculosus*', this volume, pp. 53–61.
20. Ganning, B., 'Discussion', this volume, p. 88–9.
21. Perkins, E. J., Some effects of detergents in the marine environment, *Chem. Ind.*, 14–22 (1970).
22. Crapp, G. B., 'Laboratory experiments with emulsifiers', in *The Ecological Effects of Oil Pollution on Littoral Communities* (ed. E. B. Cowell), Institute of Petroleum, London, 1971.
23. Crapp, G. B., 'The Biological Consequences of Emulsifier Cleansing', in *The Ecological Effects of Oil Pollution on Littoral Communities* (ed. E. B. Cowell), Institute of Petroleum, London, 1971.
24. Baker, J. M., 'The Effects of a Single Oil Spillage', in *The Ecological Effects of Oil Pollution on Littoral Communities* (ed. E. B. Cowell), Institute of Petroleum, London, 1971.
25. Baker, J. M., 'Successive Spillages', in *The Ecological Effects of Oil Pollution on Littoral Communities* (ed. E. B. Cowell), Institute of Petroleum, London, 1971.
26. Smith, J. E. (ed.), '*Torrey Canyon' Pollution and Marine Life*, Cambridge Univ. Press, 1968.
27. Perkins, E. J., The toxicity of oil emulsifiers to some inshore fauna, *Field Stud.*, **2** (suppl.), 81–90 (1968).
28. Simpson, A. C., Oil emulsifiers and commercial shellfish, *Field Stud.*, **2** (suppl.), 91–8 (1968).
29. Bryan, G. W., The effects of oil-spill removers ('detergents') on the gastropod *Nucella lapillus* on a rocky shore and in the laboratory, *J. Mar. Biol. Ass. U.K.*, **49**, 1067–92 (1969).
30. Crapp, G. B., 'Field Experiments with Oil and Emulsifiers', in *The Ecological Effects of Oil Pollution on Littoral Communities* (ed. E. B. Cowell), Institute of Petroleum, London, 1971.
31. Baker, J. M., 'Effects of Cleaning', in *The Ecological Effects of Oil Pollution on Littoral Communities* (ed. E. B. Cowell), Institute of Petroleum, London, 1971.

Index

143